Biosystems & Biorobotics

Volume 29

Series Editor

Eugenio Guglielmelli, Laboratory of Biomedical Robotics, Campus Bio-Medico University of Rome, Rome, Italy

The BIOSYSTEMS & BIOROBOTICS (BioSysRob) series publishes the latest research developments in three main areas: 1) understanding biological systems from a bioengineering point of view, i.e. the study of biosystems by exploiting engineering methods and tools to unveil their functioning principles and unrivalled performance; 2) design and development of biologically inspired machines and systems to be used for different purposes and in a variety of application contexts. In particular, the series welcomes contributions on novel design approaches, methods and tools as well as case studies on specific bio-inspired systems; 3) design and developments of nano-, micro-, macro- devices and systems for biomedical applications, i.e. technologies that can improve modern healthcare and welfare by enabling novel solutions for prevention, diagnosis, surgery, prosthetics, rehabilitation and independent living. On one side, the series focuses on recent methods and technologies which allow multi-scale, multi-physics, high-resolution analysis and modeling of biological systems. A special emphasis on this side is given to the use of mechatronic and robotic systems as a tool for basic research in biology. On the other side, the series authoritatively reports on current theoretical and experimental challenges and developments related to the "biomechatronic" design of novel biorobotic machines. A special emphasis on this side is given to human-machine interaction and interfacing, and also to the ethical and social implications of this emerging research area, as key challenges for the acceptability and sustainability of biorobotics technology. The main target of the series are engineers interested in biology and medicine, and specifically bioengineers and bioroboticists. Volume published in the series comprise monographs, edited volumes, lecture notes, as well as selected conference proceedings and PhD theses. The series also publishes books purposely devoted to support education in bioengineering, biomedical engineering, biomechatronics and biorobotics at graduate and post-graduate levels.

Indexed by SCOPUS, WTI Frankfurt eG, SCImago

More information about this series at https://link.springer.com/bookseries/10421

Raphael Zufferey · Robert Siddall ·
Sophie F. Armanini · Mirko Kovac

Between Sea and Sky: Aerial Aquatic Locomotion in Miniature Robots

Springer

Raphael Zufferey
Imperial College London
London, UK

École Polytechnique Fédérale de Lausanne
Lausanne, Switzerland

Robert Siddall
Imperial College London
London, UK

University of Surrey
Surrey, UK

Sophie F. Armanini
Imperial College London
London, UK

Technical University of Munich
Munich, Germany

Mirko Kovac
Imperial College London
London, UK

Swiss Federal Laboratories of Material
Science and Technology
Dübendorf, Switzerland

ISSN 2195-3562 ISSN 2195-3570 (electronic)
Biosystems & Biorobotics
ISBN 978-3-030-89574-7 ISBN 978-3-030-89575-4 (eBook)
https://doi.org/10.1007/978-3-030-89575-4

This Springer imprint is published by the registered company Springer Nature Switzerland AG
The registered company address is: Gewerbestrasse 11, 6330 Cham, Switzerland

Preface

This book combines eight years of research in Aerial-Aquatic robotics and reflects upon the progress made in this new domain from its beginnings at the Aerial Robotics Lab of Imperial College London. Through the years, multiple robotic systems to address hybrid aerial-aquatic locomotion methods have been explored by the authors of this book. The domain has now well matured, with its many varied approaches and concepts, warranting its publication in a book format.

We have structured this work to inform robotics researchers about the interest, challenges, considerations and solutions of this emerging field. The strong emphasis on bio-inspiration is highly relevant for the future of robots operating in challenging terrain and conditions. We hope that the book will also serve as a reference to students and educators in mobile robotics.

This book is split into two parts. Part I discusses the challenges of aerial-aquatic locomotion and goes on to describe existing biological solutions, e.g., animals, and current synthetic solutions, i.e., drone technologies. It ends on a short physics overview, intended as a refresher of the relevant theory at play in flight, underwater and in between mediums. Part II presents multiple robotic designs and prototypes which solve the aerial-aquatic challenge in a variety of ways, as depicted in Fig. 1. This will give the reader a demonstration and understanding of the different technologies underlying this exciting field, including a discussion of practical elements of aerial-aquatic robotics which will guide fellow roboticists in prototyping new robots of this type. The layout of the book is shown schematically in Fig. 2.

Part I of this book provides a comprehensive perspective of aerial-aquatic locomotion in both nature and science. Chapters 1 and 2 explain the concept of aerial-aquatic mobility, the opportunities and potential applications this new type of motion may lead to. Chapter 3 presents an overview of the animals capable of aerial-aquatic locomotion, based on [1]. In Chap. 4 the work continues with a summary of engineering efforts to date, and a discussion of the potential advantages that moving between air and water affords practical mobile robots.

In Chap. 5, we introduce the physical and mathematical aspects of the multi-modal locomotion problem. This includes a discussion of the basic physics involved, including the changing role of different physical effects (fluids, structural mechanics,

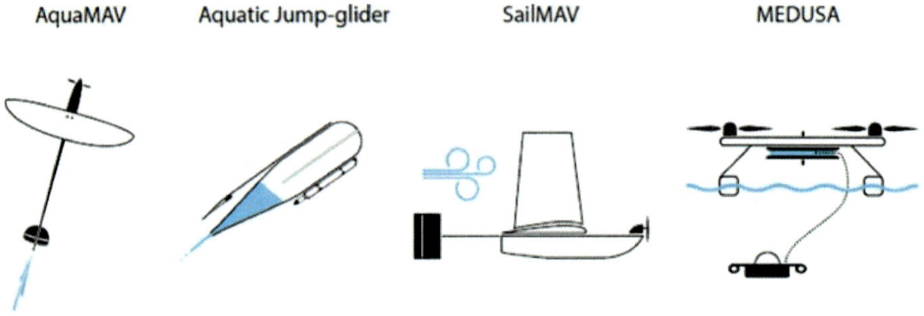

Fig. 1 Overview of the experimental robots presented in Part II of this book [1–4]. Each of these four prototypes will be discussed in detail, covering theoretical modeling, manufacturing, indoor and outdoor flight testing

power provision), and of the obstacles and opportunities of miniaturization for both robots and animals.

Part II dives deeper into the engineering and robotic development side. With a technical baseline established, the book then examines several different case studies in detail, with each chapter representing some of the latest research in mobile robotics.

Chapter 6 presents a water escape concept based on a water jet thruster, driven by the controlled release of pressurised CO_2 gas, based on [5]. In Chap. 7 we discuss its integration into a flight-capable airframe which uses deployable wings. The aquaMAV robot aerodynamic and hydrodynamic is tested in wind and water tunnel experiments, based on [6].

In Chap. 8, predictions of the trajectories and behavior of an aerial-aquatic robot are discussed based upon those measurements, based on [7]. An analytic approach to this problem is presented. It provides a detailed explanation of the forces involved, their implications on robotic design and control, and the approaches taken in the discussed robots to ensure successful air-to-water transition.

While the previous aquaMAV approach brings many benefits, mainly in terms of insensitivity to sea state, designing and implementing a vehicle that produces multiple successive water jets is extremely challenging. Chapter 9 addresses this limitation through the use of water-reactive fuel to generate a powerful jet, based on [2]. While the high-level concept of jetting is similar, the different approach used to generate the jet facilitates the execution of multiple consecutive water take-offs, thus allowing for more versatile missions to be executed.

Chapter 10 introduces a novel mechanism which enables efficient propeller-based propulsion in air and water, based on [8]. The design is presented and validated in experiments.

A unique example of aerial-aquatic locomotion is described in Chap. 11, which discusses hybrid flying-sailing locomotion on the same robot, based on [3]. Like the previous solution, this approach allows for long-duration missions, albeit only close to the water surface. Taking advantage of the wind allows for water surface locomotion with minimal energy expenditure, while the double usage of the same surfaces as wings and sails reduces the added complexity and weight of the robotic

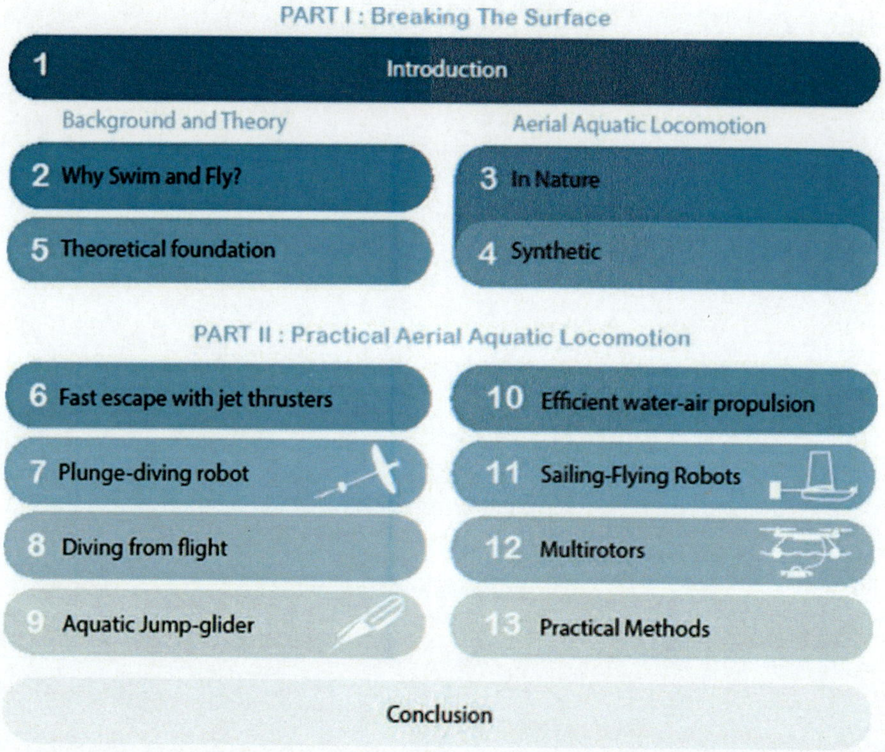

Fig. 2 Layout of this book

concept. The main drawback of this approach is its high sensitivity to the water surface conditions.

Chapter 12 describes alternatives based on multirotor robots. A novel overall approach is presented, involving two collaborating robots: a multirotor landing on the water surface and a deployable underwater pod, based on [4]. This solution allows for long-duration missions and access to narrow spaces, while also taking advantage of the ready availability of off-the-shelf multirotors, albeit at the cost of some added complexity.

Finally, Chap. 13 discusses the practical hardware considerations, including propulsion, control, communication and fabrication for reliable outdoor robots.

The book concludes with a discussion of current technological frontiers and directions for future research. This includes both active areas of enquiry and capabilities outside of current technology but well within comprehension.

London, UK

Raphael Zufferey
Robert Siddall
Sophie F. Armanini
Mirko Kovac

Acknowledgements

This book represents thousands of hours of experimental and theoretical robotics work, combining contributions from dozens of people who continue to advance engineering science around the world. In no particular order (and in a far from exhaustive description), the authors would like to acknowledge the contributions of several individuals to the work shown in the pages that follow.

Alejandro Ortega Ancel was a vital part of many of the experiments described (Chaps. 9 and 11), and an essential source of aerodynamic insight. The same is true of **André Farinha** (Chaps. 9, 11 and 12), who also brought his enthusiasm for outdoor work, mechanical design and a willingness to get wet (particularly useful in Chap. 12). Andre continues to develop unusual outdoor robots, and Alejandro now works as a data scientist.

It was **Grant Kennedy** who first hit upon the idea of exploiting water reactions as propulsion power source (Chap. 9), and demonstrated that an 'explosive water jet' was feasible. Grant now works on experimental aerodynamics for Formula One. **Yu Herng Tan** worked on the studies of aerial-aquatic propellers (Chap. 10), and hit upon the elegant arrangement of planetary gears that allowed passive gear change with significantly fewer clutch bearings. She continues to develop novel multimodal aircraft in Singapore.

We would like to thank **Celia Raposo** and **Ion Berasaluce** for their outstanding work on the SailMAV concept (Chap. 11), elaborating analytical models, physical prototypes and advanced experiments. Their engagement and relentless efforts have made this project possible. We acknowledge **Diego Debruyn** who passionately developed most of the MEDUSA robot (Chap. 12) from mechanical design to software implementation. Thanks also to **Crystal Winston** for her valuable contributions especially to the MEDUSA project.

Almost every aspect of the practical work in this book was supported in some way by the staff at the Imperial College Aeronautics Department. The authors would like to say a particular thank you to **Roland Hutchins** for his boundless patience with our unusual requests.

Finally, big thanks to **Ben Goldberg**, **Lisa Buche**, **Matěj Karásek** and **Kimberly McGuire** for their valuable insights and comments on this manuscript.

Contents

Acronyms

AC	Alternative Current
AOA	Angle Of Attack
ASV	Autonomous Surface Vehicle
AUV	Autonomous Underwater Vehicle
AWT	Air-to-water transition
BLE	Bluetooth Low Energy 4.0
BLOS	Beyond Line Of Sight
CFD	Computer Fluid Dynamics
CNC	Computer Numerical Control
DC	Direct Current
EKF	Extended Kalman Filter
ESC	Electronic Speed Controller
I2C	Inter Integrated Circuit
IMU	Inertial Measurement Unit
LCD	Liquid Crystal Display
MAV	Micro Aerial Vehicle
MOSFET	Metal Oxide Semiconductor Field Effect Transistor
PCB	Printed Circuit Board
PID	Proportional Integral Derivative Controller
PMMA	Poly(methyl methacrylate) (known as acrylic)
pre-preg	Pre-impregnated fabric
RTOS	Real Time Operating System
SMA	Shape Memory Alloy
SMD	Surface Mounted Device
TRL	Technology Readiness Level
UART	Universal Asynchronous Receiver/Transmitter
uC	Micro-controller
uORB	uObject Request Broker
VTOL	Vertical Take-Off and Landing
WAT	Water-to-air transition

List of Tables

Part I
Breaking the Surface

Chapter 1
Breaking the Surface

Most animals use different forms of locomotion to move through a varied environment. This allows them to adapt to find food, escape threats or migrate, while minimising their energetic cost of locomotion. To do so, animals must use the same locomotor modules to perform specialised tasks that often have opposed requirements. For example, an animal diving into the water to hunt requires a structure that is as lightweight as possible for efficient flight, whilst still being structurally strong when impacting the water's surface. The complexity of biological systems makes them difficult to replicate artificially, but meeting the demands of two modes of locomotion in a single robotic system has been done. Bioinspired robots have been presented that can both fly and move on land, either by jumping or by walking [9–12]. Other robots take cues from amphibious locomotion in nature and can move on land and in water [13–15].

These systems demonstrate both the feasibility and the efficacy of multimodal robotics and outperform robots with only one locomotion mode, in particular with regard to the traversability of natural terrain. The vehicles have together covered motion in water, land and air, but movement in both air and water has received considerably less research attention, yet they present a large application potential. For example, a robot capable of aerial-aquatic locomotion could play an effective role in disaster relief after floods or tsunamis, where the presence of water and debris will severely limit the operation of conventional vehicles. The potential of robotic systems for use in emergencies was highlighted by recent events such as the Fukushima nuclear accident [16], the response to Hurricane Wilma [17], and the Deepwater Horizon explosion. Section 2.2 goes into further detail into the opportunities and applications of aerial-aquatic robots.

R. Zufferey et al., *Between Sea and Sky: Aerial Aquatic Locomotion in Miniature Robots*, Biosystems & Biorobotics 29, https://doi.org/10.1007/978-3-030-89575-4_1

In these situations, a Micro Aerial Vehicle (MAV)[1] can travel rapidly over long distances to a target and return aerial footage, but would risk a difficult and dangerous takeoff/landing on the water or enter constrained areas such as flooded buildings. On the other side of the spectrum, a purely aquatic robot generally moves more slowly, is not able to pass over obstacles in the water, and is unable to gain the broad aerial perspective of a flying robot. Having a robot that can travel in air, move underwater and transition back to flight would greatly improve search and rescue capabilities.

An aerial-aquatic vehicle would also have great potential as a sample collection or observation tool for oceanography research, a field in which air-launched underwater vehicles are already sought after [18, 19]. Research has already demonstrated the effectiveness of combining aerial and aquatic robotic operation by co-ordinating separate systems as a team for the monitoring of coral reefs [20], with an MAV using aerial video footage to guide and augment close observation by an aquatic vehicle. Murphy et al. [17] also found that addition of an MAV provided a more effective communication relay with the surface vehicle, a topology also used in the MEDUSA robot presented in Chap. 12.

Mission Phase	Design Challenges
Dry Flight (Fig. 1.1A)	Low energy flight, low weight, robustness and stability
Water Entry (Fig. 1.1B)	Accurate sensing and control (soft landing) or high impact loads (direct dive)
Submerged (Fig. 1.1C)	Low buoyancy, provision of waterproofing, efficient propulsion and autonomous navigation
Surface Locomotion (Fig. 1.1D)	High buoyancy, light waterproofing, wave-resistance, efficient propulsion
Water Exit (Fig. 1.1E)	High power density (direct launch) or low surface drag (taxiing takeoff)
Wet Flight	Water shedding

1.1 Challenges

Hybridising flying and swimming presents several key design challenges. To analyse them, it is helpful to examine a robot's mission in the context of distinct mission phases (Fig. 1.1): First, the robot flies to a target, then must transition from air to water. After transition, it must spend a period moving beneath or at the surface collecting water samples, aquatic measurements or video footage, before transitioning back to flight from beneath the water.

[1] We use the term MAV in this book to refer to small, sub-kilogram, unmanned aerial robots. This term is widely used in the literature to cover robots that are technically much bigger than the micrometer scale.

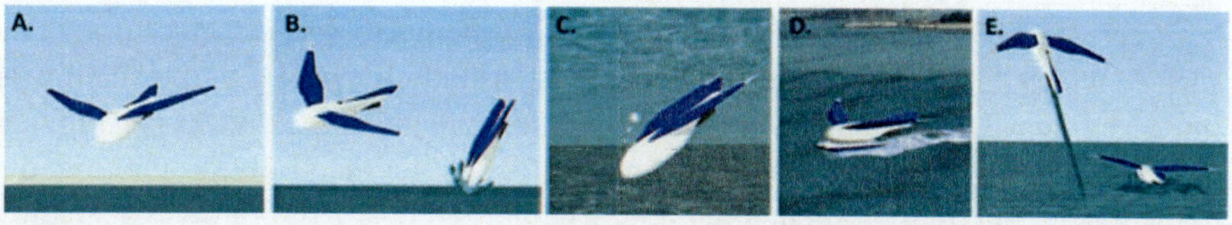

Fig. 1.1 Every stage of an aerial-aquatic mission poses different key challenges for a generic aerial-aquatic robot: (**A**) Dry flight, (**B**) Water entry, (**C**) Submerged movement, (**D**) Surface movement and (**E**) Water exit

1.1.1 Dry Flight

Powered flight is generally power-intensive, and the key challenge during this phase is the minimisation of energy consumption. This is broadly achieved by maximising the aerodynamic efficiency of the vehicle, which becomes more difficult at the small scale. Micro Aerial Vehicles typically operate in the Reynolds number range 10^4–10^5, and at these low flight Reynolds numbers the aerodynamic efficiency of flight is reduced due to increased viscous effects. At small scales, air perturbations will also have a much more significant effect on a vehicle [21], creating challenges in stability and robustness. Furthermore, the design constraints of flight must be satisfied simultaneously with those imposed by the need to swim. These additional constraints come from the need for dual propulsion modes, a stronger superstructure and waterproof design, all of which add to the flight mass.

Materials considerations for the wings are essential not only for the minimisation of structural mass. Selecting materials to add flexibility to the wing structure, for example, also offers many enhancements for low Reynolds number vehicles, from aerodynamic improvements, obtained by allowing both passive and active adapting of wing morphology [22], to structural benefits, obtained by improving durability [23].

1.1.2 Water Entry and Landing

The challenges associated with entering the water depend on the strategy for transition. One proposed approach in this book is plunge diving directly through the surface which carries over flight momentum into the water, and can allow depths to be reached rapidly. However, in order to do so a vehicle must be as slender and streamlined as possible to minimise impact loads. Diving impact loads have been found to scale approximately with the square of impact velocity (see Chap. 5), and result largely from the accrual of fluid inertia during impact. The added mass will scale with the body cross sectional diameter to the third power (an oversimplification, but one that highlights the importance of minimising cross section during impact). This makes effective deep diving highly impractical for fixed-wing aircraft whose wings create a large frontal area, and no animal attempts to dive directly through the surface without folding or morphing its wings. A body impacting at too shallow an impact angle will experience a large bending moment, and if the dive has insufficient momentum it may rebound from the water surface rather than fully penetrating, due to the impact pitching moment and rapid deceleration [24, 25]. For dives at shallow angles, the eccentricity of the impact load also has a tendency to flip a vehicle, which is a significant problem for soft floatplane landings.

Alternatively, a vehicle can perform a soft landing, whether by gradual descent or deep stall just above the surface. This approach reduces structural requirements and may mean folding wings are not necessary, but an additional system for underwater propulsion is then needed, and the robot cannot benefit from flight momentum when diving into water.

A vehicle that does not require depth operation may stay at the surface. While this limits the application spectrum to surface sensing and data collection, it eases the constraints of underwater locomotion. Most notably, the buoyancy of the body can remain high at all times, a characteristic that is compatible with flight and surface operation. This gradual landing strategy requires more accurate control capability and reliable sensing of attitude and altitude, which adds complexity.

1.1.3 Submerged Movement

Once the vehicle is in the water, it must handle resistance from both buoyancy and drag. The latter in particular will be several orders of magnitude higher compared to the drag encountered during aerial motion at the same speed, due to the increased density and viscosity in water. The benefits of reducing buoyancy for efficient swimming must be traded against low weight for efficient flying. For non-zero buoyancy, the centre of volume as well as the vehicle's hydrodynamics will determine its stability, which will be important if a plunge dive and passive, buoyancy-driven ascent is desired. As the robot travels deeper, hydrostatic pressure loads will also become significant, and place harsher requirements on the robot's structure and the waterproofing of electronic components. The limited propagation of electromagnetic waves in water [26] will also make remote control of robots highly difficult, forcing the vehicle to instead rely on some degree of autonomy.

Finally, the most pernicious consequence of underwater operation is the risk to electronics from water ingress. This is especially challenging where weight is constrained, as the thick, rigid gasket seals often found in underwater vehicles are too heavy. Electrical shorts and corrosion from water can cause significant delays to prototyping and development and should not be underestimated.

1.1.4 Surface Movement

As soon as the vehicle contacts the water surface, it will either cross the interface or remain at the surface. In the latter case, the vehicle needs to be supported by sufficient buoyancy, surface tension (at or below gram-scale) or lifting underwater wings (i.e. foils). Then, the vehicle can simply remain at the surface without any means to move horizontally. This limits the applicability of the robot to a spot measure, essentially acting as a flying buoy. In order to move at the surface, the vehicle requires some propulsive method. In this scenario, the energy source can be either on-board or provided by the environment. The on-board propulsive power can come from the flight propellers (albeit at low efficiencies) or from additional power sources (chemical, aquatic electric propeller etc.). The environment can potentially provide energy for water surface locomotion in the form of wind, wave and aquatic current energy.

1.1.5 Water Exit

The key challenge for fixed-wing robots is building up speed for flight. However, most MAVs in regular use do not depend on their own power to take off, but instead rely on gravity or on being thrown in order to generate their initial flight speed. This difficulty is only compounded by aquatic locomotion, where hydrodynamic drag limits the speeds achievable [27], especially at the surface, where additional effects such as wave and spray drag can increase resistive forces by 5 times their value when fully submerged [28]. Wave drag is caused by the formation of a wave pattern around a body, the wavelength of which increases with speed. Typically, waves are first formed at the front of the ship (i.e. the bow), and are known as bow waves. As the wavelength exceeds the body's length, the stern (rear end of the ship) begins to sit in the wave trough, and the vehicle is forced to climb its own bow wave, dramatically increasing power requirements until hydroplaning conditions are achieved.

Furthermore, the take-off power requirements will be significantly increased by the added weight of water [29], which must be shed as quickly as possible. Wind-driven surfaces waves can also be a significant problem for small scale aerial-aquatic vehicles; just as the relative size of gust perturbations increases for micro scale flight, so does the relative wave height for micro scale surface swimming. Waves passing across the vehicle can immerse lifting surfaces and cause the vehicle to crash, and for a taxiing takeoff, the vehicle may then have to move in the same direction as wave propagation, which will require additional sensors.

Some aerial-aquatic vehicles are Vertical Take-Off and Landing (VTOL) capable. This encompasses all robots from standard multi-rotors to hybrid winged and lifting propeller configuration. The VTOL flight capacity removes the requirement to reach high speed to generate lift. However, this technology is challenging from the water, due to the propeller clearance necessary. Surface state will have a large impact on this capability, often hindering take-off or even overthrowing the system. Robustness in varying sea state will be a key challenge necessitating innovative solutions.

1.1.6 Wet Flight

Once the vehicle has left the water with sufficient velocity, it then needs to return to efficient flight. To do so it will be important to shed any water weight picked up during swimming. The amount retained surface water will depend on the surface's properties, but will broadly scale with surface area. Because surface area to volume ratio decreases with size, this may represent a significant barrier to the miniaturisation of aerial-aquatic robots.

Flight trajectory planning will also be important, as the ground effect can be exploited near the water surface to reduce induced drag, which has been shown to alter maximum range gliding flight paths [30]. However, water waves and spray will restrict the minimum flying height and affect the local airflow near the surface, creating flight perturbations, so flight must be robust in order to do this.

1.2 Multimodal Mobility and the Outdoor Environment

While aerial-aquatic robotics is a fast-expanding research area, an overwhelming majority of experimental results occur in controlled conditions. Yet, the draw towards these hybrid systems is mainly their ability to operate in normally inaccessible areas of the natural environment. To realise their potential, aerial-aquatic vehicles need to bridge the uncertainty gap between laboratory and outdoor conditions. The small size of Micro Aerial Vehicles brings appreciable benefits in terms of transport, deployment and cost. But it also magnifies the effect of external perturbations, and increases the challenge of operating in unstructured environments, as discussed below.

Air The ambient fluid motion in air (i.e. wind) is often unpredictable and perturbed, with wind velocities often exceeding typical UAV flight speeds. Even in mild conditions a so-called "light breeze" on the Beaufort scale involves winds of up to 11 km/h. Even this supposedly gentle wind speed may overwhelm slow-flying robots and make certain flight trajectories infeasible. Generally, aerial-aquatic vehicles should be able to fly at speeds above a threshold wind speed or transition to the water if the air conditions do not permit flight. Natural air convection is typically turbulent, and wind gusts can present a considerable challenge to the stability and control of aircraft, especially at small scales, where flight speeds are typically lower.

Underwater Outdoor aquatic environments likewise present challenges. Underwater environments may change drastically from a clear lab pool to an ocean, river or lake—and apart from certain stagnant ponds, every outdoor body of water will be in motion. These currents will make robots drift and need to be countered to maintain position control, which often requires an external static reference (e.g. the seafloor). An additional considerable challenge is communication: at depth, radio communication is not possible and navigation needs to be autonomous, relying on additional sensors.

Surface Where the air and water meet, so too do their ambient motions, with wind-propelled waves interacting with water currents. In the same "light breeze" Beaufort scale conditions mentioned previously, waves of up to 20 cm can be expected, and for small aircraft with low weights and large surface areas even slight surface motion represents a considerable challenge. The impact of waves can be mitigated through impulsive actuation (see Chap. 6), opting to pierce through the surface rather than travel along it. However, some form of surface operation capability may be necessary or even beneficial to certain types of missions, offering vehicles a location where they can loiter indefinitely with access to solar energy as well as radio communication covering a majority of the earth's surface. For locomotion on the surface, additional locomotion penalties such as wave and spray drag must also be accounted for, and a body shape optimised for piercing the waves may be suboptimal below the surface. Additionally, water surface locomotion becomes increasingly challenging at very small scales, where the size grain hypothesis implies that wave height becomes increasingly large relative to body size.

1.3 Aerial-Aquatic Robots Presented in This Book

This book presents several robotic solutions allowing for operation in air and in water as well as for inter-medium transitioning. Four examples of different types of robots are discussed in particular detail: all of these were built to validate and compare different aerial-aquatic locomotion methods. The robots (i.e. the solutions) are characterised by key differences in their distinct *aerial-aquatic cycle*, which we briefly outline in this section. Note that the following sub-figure references all relate to Fig. 1.2.

Firstly, we present a folding-wing flyer, which combines fixed-wing aerial locomotion and propeller-driven underwater locomotion, and is capable of diving into the water. The 'AquaMAV' [5, 6] showcases a dynamic air-to-water transition approach based on a plunge-diving manoeuvre (**a**), which is less sensitive to the sea state compared to landing on the water surface. From the air, the two flat-plate wings fold back, initiating a dive. The robot hits the water nose first, with the propellers off and folded. The wing morphing reduces the impact forces on the robot and the drag in water, permitting a dive to several meters of depth. An unactuated gearbox (**b**) reduces the rotation speed of the propeller to 1:15 [8], resulting in more effective underwater motion. The AquaMAV impulsively jets out of the water (**c**). At this stage, the wings are extended, and propelled flight carries the vehicle for up to 5 km.

Secondly, we discuss the water-reactive fuel jump-glider. This is also a fixed-wing vehicle, similar to the AquaMAV in size, and with a weight of 160 g (**j**). The

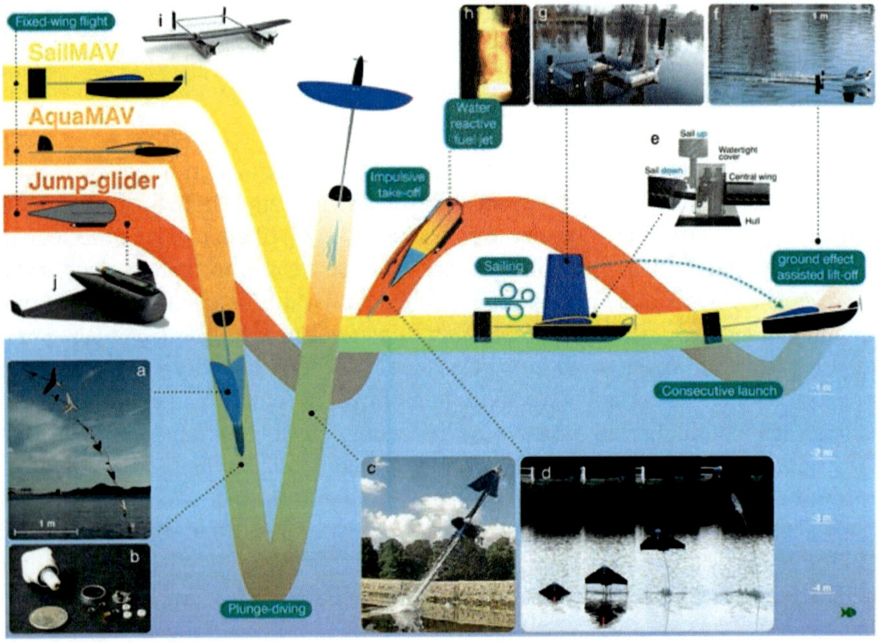

Fig. 1.2 Visual comparison of impulsive and lift-off water exit methods

particularity of this vehicle, however, it that the provision of water-reactive fuel on board allows for multiple consecutive water exits (i.e. jumps out of the water). The impulsive power required for the aforementioned exits is generated through the combustion of acetylene (**h**), which creates a powerful water jet. Peak pressures of 7.8 bars accelerate the robot out of the water (**d**), where it then transitions into a gliding trajectory.

Thirdly, we demonstrate a different form of multi-modal locomotion, with a robot combining sailing and flying capabilities. The 'SailMAV' flies as a fixed-wing platform (**i**) with a 0.9 m three-section wing and a flight distance of 7 km. The carbon fibre asymmetric central wing provides high lift and stiffness, while the extremities (labeled as *sails*) have two degrees of freedom, i.e. angle of attack and folding angle. Once landed on the water surface, it leverages wind energy for passive locomotion (**g**). During this phase, the sails are in an upright position, locked in place by a folding mechanism (**e**). Sailing is autonomously handled by a modified flight controller. By lowering the sails to a horizontal position, the SailMAV reverts to flight mode. At 3 m/s velocity, the robot begins to hydroplane thanks to specially designed hulls, at which point the drag decreases sharply. Take-off occurs at a low velocity through the help of ground effect (**f**).

Lastly, we present a dual robot, MEDUSA, consisting of a combination of a multirotor, capable of landing and floating on the water surface, and a retractable tethered underwater pod. Breaking away from the fixed-wing theme overarching the rest of the research presented in this book, this solution offers high maneuvrability in flight, thanks to its multirotor configuration, as well as agile underwater motion thanks to the separate pod.

In the second part of this book, these robots will be discussed in further detail. Together, they exemplify a wide variety of different locomotion and operation concepts, and illustrate the key challenges involved in aerial-aquatic robotics.

Chapter 2
Why Swim and Fly?

2.1 Motivation and Objectives

We live on a water-covered planet that is facing rapid change, both globally and locally, due to a combination of human behaviour and natural phenomena [31]. Understanding these changes requires in-depth scientific understanding of our environment. Key to enabling this is the fast, accurate and repeated provision of extensive physical data. However, obtaining dependable geospatial data is itself frequently a challenge, requiring new sensing approaches or considerable adjustments to existing methods. In fact, in many cases, existing approaches are either not convenient or entirely inapplicable for the necessary data generation.

A particularly challenging class of information to obtain is aquatic data, which serves to estimate a large variety of parameters [32]. Aquatic environments can be extremely difficult to access, particularly environmentally relevant locations such as arctic regions and coral reefs. In addition, spacial and temporal resolution is low with manual measurements [33]. These current measurement approaches often involve long-duration mission and high costs, and are further constrained by underwater communication limitations. The development of small and affordable robotic vehicles capable of performing aquatic measurements could play a significant role in sourcing data for essential environmental models, as well as supporting a fast and effective response to local aquatic emergencies such as floods, storms or industrial spills. Numerous specific applications would thus benefit from the availability of such robots, further detailed in Sect. 2.2.

Clearly, the ability to transition to flight would vastly extend the potential and capabilities of AUVs and ASVs, allowing for extremely fast deployment, rapid coverage of extensive distances, overflying of obstacles and autonomous transition between different bodies of water. In addition, the possibility of landing on the ground would eliminate the need for a boat or port to recover such robots.

R. Zufferey et al., *Between Sea and Sky: Aerial Aquatic Locomotion in Miniature Robots*, Biosystems & Biorobotics 29, https://doi.org/10.1007/978-3-030-89575-4_2

2.2 Applications and Opportunities

2.2.1 Remote Sensing

Remote sensing, i.e. the observation of Earth from space, allows us to detect and attribute spatial and temporal trends in ecological properties, such as sea surface temperature or photosynthesis rate [34]. This information is crucial as it provides a regular and holistic view of the global environment, which directly influences climate change policies. Satellites can process reflected electromagnetic radiation from almost any region on Earth, regardless of accessibility at the surface. However, external calibrations are needed to link satellite signals to properties measured on Earth, and allow for measurement accuracy to be maximised and for signals to be interpreted in physical terms. At present, the critical measurements required for satellite 'ground-truth' at sea are difficult to produce [35] and only obtainable from dedicated buoys or expensive ship-based missions. Due to the low number of clear sky buoys and ship measurements that co-occur with satellite passes, calibrating satellite measurements can take from years to decades and is a major and consistent source of uncertainty. Aerial-aquatic robots could be extremely valuable in this context. Indeed, flying robots are well-suited to map 'pixel-sized' areas, and drones that could land and move on or under the water could provide similar mappings for ocean surfaces. Affordable hybrid robots could perform the external calibrations required in remote sensing applications far more quickly, efficiently and cost effectively than either ships or buoys are currently able to.

2.2.2 Marine Conservation

Aerial-aquatic technology could also provide enhanced capabilities to marine conservation efforts throughout the planet. The Maldives, for instance, is composed of 1200 islands spread over an area larger than the United Kingdom, and would benefit immensely from small, autonomous robots for monitoring marine resources, marine tourism and fisheries. Drones are already being used, for example, during the Manta Ray season to monitor Hanifaru bay in the Baa atoll UNESCO Biosphere reserve, however their use is limited in scope owing to their inability to interact with the water and their short range so far. Aerial-aquatic platforms would supply the currently lacking underwater-inbound data. For example, underwater noise can be monitored to estimate the number of individuals and species. The MEDUSA platform (see Chap. 12), for instance, is ideally suited for this application category due to its underwater mobile pod which can adequately reach shallow reefs and plants, and manoeuvre among these.

2.2.3 Micro-biology and Micro-plastic Analysis

Micro-plastic is becoming an increasingly urgent issue. Our seas are being filled with small pieces of plastic originating from the breaking down of larger parts, typically ill-managed waste. In order to tackle this issue, we need a better understanding of the flows, sources and accumulation sites, according to National Geographic [36]. Micro-plastics are also present in fresh water systems and rivers, where monitoring is also required [37].

Micro-plastic pollution is measured by counting the amount of particles of given sizes in a water sample of fixed volume. Such a water sample could be taken and transported back by an aerial-aquatic robot in an automated manner. This would allow an increase in sample frequency and largely reduce the expenses linked to manned boat operations.

However, the sample weight that could be carried by a small aerial platform will be limited, reducing the quality of the subsequent analysis. Further particles can be collected through concentration. A mechanical micro-filtering system would collect particulates and return with minimal amounts of water. The quantity of filtered water would therefore only depend upon the collection time, bounded by the pump flow. In addition to micro-plastics, a micro-filter aerial-aquatic system could be expanded for use in micro-biology studies, e.g. for plankton concentration measures. In a similar vein, sampling of micro-organisms and chlorophyll can also be used to study anthropogenic effects such as climate changes.

2.2.4 Pollution Monitoring

Pollutant levels rise in rivers as the waters approach the sea. The monitoring of these pollutants is difficult due to the frequent sampling required. The methodology is tedious, and involves a large amount of field work. Aerial-aquatic technology would be helpful in providing data regarding the water quality of polluted rivers (and other water bodies). The use of this technology would facilitate the implementation of such monitoring projects and provide a more rigorous framework. A studied application scenario is the Bengal area where the waters of the Himalayas are drained into the sea, transporting sediments but also heavily polluted. The flow of the river complicates the application of small robotic vehicles and therefore only a robust, single-launch type aerial-aquatic robot could perform a successful mission in such a scenario. Indeed, jet-powered water exit capability renders robots relatively insensitive to water surface roughness and currents. Within Europe, research programs seek to improve the quality of the rivers [38], which would also benefit from aerial-aquatic technology.

2.2.5 *Arctic Research*

At an entirely different latitude, aerial-aquatic robots are likely to prove highly beneficial for arctic research, where aerial-aquatic technology could support a large variety of activities. This includes (but is not limited to) sampling remote lakes, patrolling glacier toes, studying melt ponds formed on sea ice, navigating in the vicinity of dangerous icebergs, investigating sediment-rich glacial flows and understanding the formation of polynyas. The polar regions are a threatened and irreplaceable information resource, preserving millennia of unique environmental data. They are also hostile and dangerous environments where data collection costs time and resources, and multi-modal robotics offers a chance to greatly improve data collection. In addition, glacial flows transport a rich amount of sediments, poorly understood so far due to the difficulty of taking samples by hand. An aerial-aquatic robotic solution would not only permit to obtain samples autonomously, it would also drastically lower the risk that researchers undergo in such missions.

There are many other limitations as to where a vessel can go. Unreachable areas include shallow water, shore line, protected areas, ice field surrounded regions, fjords, small lakes, rivers and more. Further limitations are dangerous areas where manned access would be impossible, most importantly crater lakes, edge of glacier terminals, iceberg vicinity, melt ponds.

2.2.6 *Marine Wildlife*

Our marine life is constantly changing and evolving. Monitoring technologies could vastly improve our understanding of this evolution and enable us to act appropriately. While aerial imaging, whether from satellites, planes or locally from drones already provides visual information about the surface activity, underwater life is harder to record.

Aerial-aquatic robots could potentially transmit downward-facing aerial images, which can be employed for measurements or for subject tracking. Once landed at the water surface, the cameras would give images of the underwater world. Fish species could be identified from these images, as well as sea-weed density. In addition, this technology could provide a way to ground-truth purely aerial images. Indeed, the underwater image could serve for data-labelling purposes for the aerial picture of the same area.

Overall, hybrid robots show potential to track our marine life, whether plants, small fish or whales. Such vehicles will require high wave resistance as well as reliable waterproofing due to the corrosive salty environment. Additionally, long surface operation time would be a significant advantage to study slow-moving subjects. Passively-powered robots, such as the sailing-flying platform (see Sect. 11.8) would be a good candidate for this mission profile.

2.2.7 *Offshore Platform Maintenance*

Aerial-aquatic robots will also be a valuable tool in a range of industrial applications. A domain where such robots could prove particularly advantageous is the offshore sector. Routine maintenance and inspection operations of (partially) submerged structures are often challenging, especially in coastal regions and splash zones, difficult to reach using large water surface vehicles. Indeed some areas, such as the spaces underneath oil rigs, are not easily accessible with available robotic inspection methods (such as purely underwater robots) and generally rely on rope-access, which poses safety risks and implies high costs.

Small-scale aerial-aquatic robots can access locations dangerous or impossible to reach by other means, conduct autonomous underwater visual monitoring tasks, aided by their underwater locomotion capabilities, and return to base with new data once their mission is complete. Such process would potentially lead to higher safety, efficiency and cost-effectiveness compared to current solutions (e.g. human-dropped sensors, direct rope access, purely underwater robots).

In addition to routine underwater infrastructure inspection operations, aerial-aquatic robots would also be valuable to manage incidents (e.g. spills), by monitoring water quality, identifying harmful substances and generally informing manned responses to such incidents. Moreover, thanks to their features, aerial-aquatic robots could be an effective tool for mapping the underwater ecosystem around offshore platforms, supporting sustainable asset management and decommissioning decisions.

As an extension of the previous application example, aerial-aquatic robots could not only be used specifically in the context of offshore platform maintenance, but also more generally for any manner of application involving underwater maintenance, particularly in shallow waters.

2.2.8 *Search and Rescue*

As well as allowing for new applications or improved procedures in various research, ecology and industry domains, aerial-aquatic robots may also be of advantage for more classical missions associated with UAVs. One such example is search and rescue. Thanks to their enhanced mobility and great versatility, aerial-aquatic robots would for instance be useful for reaching water bodies and moving between different areas rapidly, while also being able to search on and under the water surface, rather than requiring a combined effort of separate aerial and aquatic vehicles. While most existing vehicles are too small to directly proffer assistance, they may nonetheless be highly efficient at searching and identifying rapidly across water bodies, thus accelerating the initial stages of a search operation.

2.2.9 *Extension of Operation Envelope of Cruises*

Current ocean research expeditions operate on large budgets (typically 20.000 £/day) and sampling/measuring is limited to the geographical position of the ship. Typical tools are lowered off the side of the vessel into the water. This implies that data is only collected along the trajectory of the cruise. Aerial-aquatic robots could serve as a shuttle to locations away from the ship's trajectory. The robots could fly from the deck of the ship, collect a measurement before flying back to the ship. Given sufficiently capable robots, the ship would not even necessarily need to stop if the flight speed of such robots exceeds the ship's velocity.

2.2.10 *Bathymetric Mapping*

One important application of aerial-aquatic robots is sea-floor mapping. Bathymetric mapping uses echo sounders to build a 3D map of the underwater environment. This technology, typically deployed from ships, is applicable to aerial-aquatic robots as the measurements can be taken from the surface. While the accuracy of the sensors that can be fitted onboard a lightweight robot certainly will not match the ship's capabilities, this technology is interesting to quickly map small underwater zones in coastal areas. Especially the ability to obtain a 3D map from a shore-launched robot would reduce the cost of the operation and allow frequent mapping of moving underwater terrain.

2.3 Aerial-Aquatic Robots as a Solution

With their inherent multi-functionality, aerial-aquatic robots would be a valuable asset in numerous challenging tasks, many of these essential to the continued increase in our understanding of the environment. Applications of such robots in an industry context are also promising. However, aerial-aquatic locomotion is highly complex and difficult to achieve in robots, especially at small scales. The following chapters of this book provide a comprehensive introduction to the concept of aerial-aquatic locomotion, starting from nature, where evolution has generated a variety of solutions to tackle this challenge. The book then moves on to discuss the latest developments in robotics, and concludes with a broad range of detailed case studies in practical aerial-aquatic locomotion.

In order to fulfil any of those missions, a robot needs to be fitted with sensing/sampling capabilities. In Sect. 13.8, we summarise sensing option for small vehicles.

Chapter 3
Aerial-Aquatic Locomotion in Nature

3.1 The Pelagic, Pleustonic and Littoral Environments

Water covers 363 million square km, or 72% of the earth's surface. The vast majority of this water is saline (96%), frozen (2%) or groundwater (1%). The $10^5\,km^3$ of surface freshwater (0.008%) is in turn concentrated almost entirely in three large great lake systems (Fig. 3.1), with a vanishing small amount of surface freshwater forming lakes and rivers. Life on Earth gathers around water, and 90% of the human population lives within 10 km of a freshwater source (though growth in the use of aquifers is changing this).

From the Earth's freshwater reserves we derive the majority of all living material, in the form of terrestrial plants (Fig. 3.1). The land habitats of the earth, with their greater insolation and 'architectural diversity' have become the earth's reservoir of biomass. The geographic isolation created by continents, topography and extremes of temperature create evolutionary laboratories that drive evolutionary change.

By comparison, the ocean is a refrigerator, with a far more stable environment over evolutionary time. However, this vast, fully connected habitat contains the bulk of animal life, predominantly in the form of fish and marine arthropods (Fig. 3.1). As a result, where the productive land meets the rapacious sea we find concentrations of life, and coastal (littoral) waters have a disproportionate role in marine ecology. Only 0.5% of ocean volume is coastal, yet their ecosystems represent 90% of annual marine biomass creation.

The landmasses of the world are surrounded by a submerged continental shelf with a consistent depth of around 140 m (this consistency is a relic of sea level falling during a recent ice age). Somewhat conveniently, light penetration into the ocean varies between 50 and 200 m, depending on turbidity and insolation. This is what makes this region abundant; the combination of nutrients flowing in from the land and sunlight which penetrates to the seafloor (benthos). This allows energetic predator-prey interactions driven by visual pursuit, and results in faster and more active animals. It is these animals which are of particular interest to engineers seeking locomotion adaptations.

R. Zufferey et al., *Between Sea and Sky: Aerial Aquatic Locomotion in Miniature Robots*, Biosystems & Biorobotics 29, https://doi.org/10.1007/978-3-030-89575-4_3

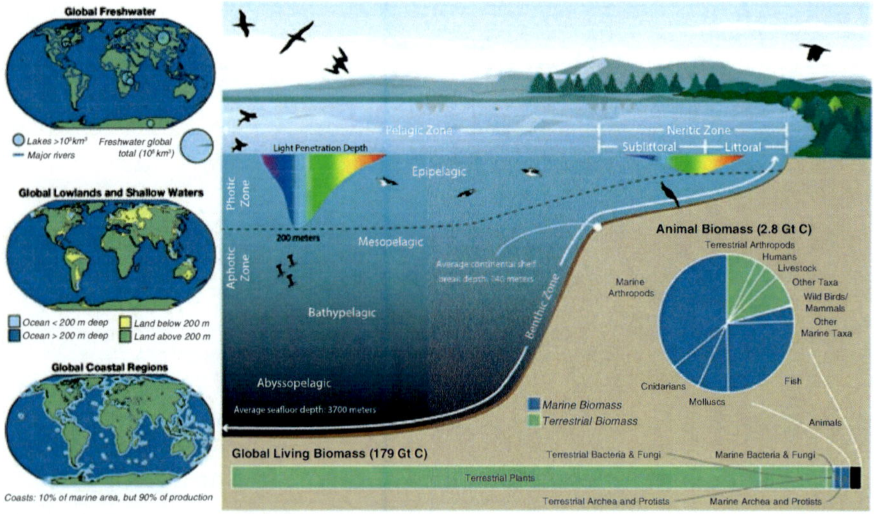

Fig. 3.1 The interface of the aerial and aquatic ecosystems. While most biomass is terrestrial, animal life is mostly aquatic (Biomass estimates taken from [39]). The neritic zone above the submerged continental shelf, where light penetrates to the nutrient-rich benthic zone, is an important reservoir of animal life, and full of evolutionary innovations for aerial-aquatic locomotion, driven by competition between predator and prey

It is perhaps not surprising that many aerial animals take advantage of such plentiful aquatic prey. Flying animals have the advantage of range, and access to offshore hunting grounds out of reach of other terrestrial animals. A myriad of animals have evolved a seamless transition between powered locomotion in water and air, and move through both media with the efficiency required to forage effectively. Equally, many aquatic animals use the air as a means of escape from underwater predation, and spend extended periods on the wing above the surface.

For all of these animals, one of the strongest evolutionary pressures is the need to reduce the energetic cost of locomotion, while still being capable of moving through varied terrain; the resulting adaptations represent an extended and unforgiving design process. Bioinspiration then becomes a potent resource for roboticists, particularly where conventional, single-mode design strategies become less applicable. This is especially the case for the design of miniature robots, where scale makes terrain less traversable, energy and power density is limited, and transport efficiency is of paramount importance. Robotics can not only benefit from natural inspiration, but can also be used as a tool with which to understand natural locomotion (e.g. [40–42], where many unanswered questions exist.

However, bioinspiration should not typically entail the blind copying of the morphology or behaviour of natural systems, and animals face different design constraints to their robotic counterparts. Evolution itself moves towards sufficient rather than optimal design, and the evolutionary pressures faced by animals and robots are not always the same. For example, many of the features common in gliding vertebrates, such as a dorsoventrally flattened body or webbing between limbs may be used by other animals for display or disguise, rather than any aerodynamic purpose [43]. Because of this, it is important not to simply duplicate nature, but to examine and abstract key principles which guide practical engineering design.

The following sections describe some of the features that animals which move between air and water have evolved, and examines the compromises that must be made to move in both media (Table 3.1). The chapter concludes with a discussion of the animal adaptations which can inform the design of aquatic aircraft.

3.2 Swimming in Aerial Animals

One of the more successful adaptations in natural history is the feather, which has allowed warm-blooded birds to penetrate the coldest regions of the planet, and move through the air with enviable fluidity. Partially because they are able to withstand frigid ocean temperatures, a wide variety of flying birds can also move beneath the water. To takeoff again, many birds will be able to lift themselves from the water purely by flapping, but many heavier birds that have adapted for swimming must first accelerate. To avoid wave drag penalties while doing this, birds will lift themselves out of the water as they accelerate, and hydroplane on the water surface through a combination of flapping and foot propulsion [44]. This can be seen in Fig. 3.2B, in which the suppression of wave drag is made apparent by the absence of a bow wave in front of the birds. However, the large amount of spray scattered by the birds also gives an indication of how energetic a process this is. Most aquatic birds execute this kind of taxiing takeoff, with foot propulsion supplementing their flapping power. Where diving birds differ the most is the manner in which they move below the surface. Broadly speaking, underwater hunting methods in birds can be divided into plunge diving directly from flight, and pursuit diving, in which underwater locomotion is initiated from a stationary position on the water surface.

3.2.1 Plunge Diving

Gannets (genus *Morus*) have one of the more spectacular strategies for catching prey below the water surface. These birds plunge dive from the air at heights of up to 30 m, and entry speeds of up to 24 m/s [46]. To maximise their penetration depth, gannets have a long slender body, and wings that can be swept fully backward to minimise their cross section when entering the water (Fig. 3.2D). They also have subcutaneous air sac structures, pressurised through connection to the lungs. These structures streamline their bodies, cushion impact and provide some means of buoyancy control [47]. Recordings from data loggers attached to the backs of diving gannets record no measurable impact deceleration at a sampling rate of 32 Hz, as argued by Ropert-Coudert et al. [48].

Fig. 3.2 Strategies for aerial-aquatic locomotion: (**A**) Two stage transition: Flying Fish, (**B**) Hydroplaning: Ducks, (**C**) Jet propulsion: Flying Squid, (**D**) Plunge diving: Northern Gannet, (**E**) Morphing wings: Common Guillemot, (**F**) Foot propulsion: Great Diving Beetle. *Image credits: US NOAA* (**A**), *Cappi Thompson* (**B**), *Michael Bamford* (**C**), *David Tipling* (**D**), *Lock et al. [45] (Unaired BBC footage)* (**E**) *and Warren Photgraphic* (**F**)

Interestingly, a recent computational simulation of gannet water impact [49] has predicted high impact decelerations (up to 23 g for a 24 m/s dive) with force peaks lasting on the order of 0.1 s, which should be measurable by the loggers in [48]. The simulation treats the gannet as rigid, and the disparity could be accounted for to some extent by the gannet's air structures and feathers. The feathers of the gannet are also both air filled and hydrophobic, so will entrain and release an air layer around the bird's body during a dive, likely to an extent that was not captured in [49]. This has been suggested as a source of drag reduction through air lubrication in penguins [50], but it should also be noted that [49] predicts a minimal contribution to impact force from skin friction. Another consideration is the fact that a gannet will in some instances exceed the surface cavitation speed during impact [51], though well below the speed to form a natural supercavity.

While capable of some swimming, the birds generally achieve depth purely through the momentum of their dive [52]. When they have reached their desired depth, they redeploy their wings as brakes, before allowing buoyancy to propel them back to the surface. While this strategy of minimal swimming limits the birds' underwater range, it is noteworthy that compared to other diving animals the gannet has a much higher foraging success rate in terms of captured calories/hour, something necessitated by the energy required to remain airborne whilst seeking a target. These birds are more efficient flyers than most aquatic birds, with large, high aspect ratio wings for efficient loitering whilst seeking a dive target. The downside of such large

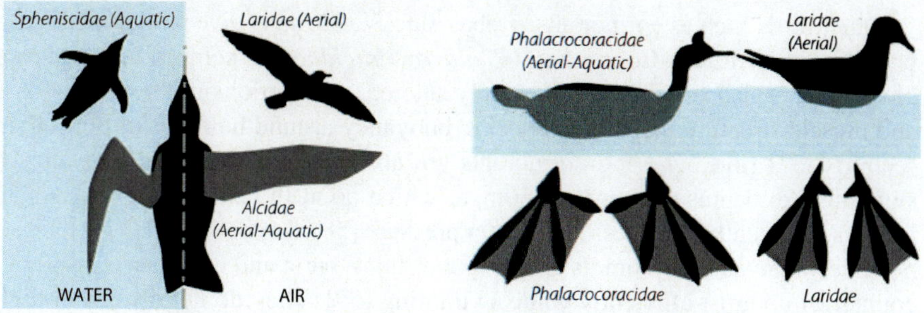

Fig. 3.3 Example adaptations in aquatic birds. Diving alcids (guillemots, murres puffins) have smaller wings relative to their size than purely aerial species, but still larger than the wings of fully aquatic birds (e.g. penguins), and so must partially retract their wings underwater to swim efficiently. Pursuit diving birds such as cormorants have much more wettable feathers, to reduce their buoyancy as well as larger webbed feet than typical surface swimming seabirds, via totipalmate dactyly

wings comes when attempting to leave the water, where the flapping stroke length is constrained. Gannets on the surface of rough seas often must wait for the water to calm before they are able to takeoff [53]. To compensate for their limited flapping power at the surface, the birds takeoff into the wind whilst holding their wings aloft to generate initial lift, and use dynamic soaring flying techniques, gaining altitude using air currents deflected by surface waves (Fig. 3.3).

Plunge diving behaviour is not limited to gannets and boobies (*Sulidae*), but is also exhibited to a less striking extent in smany other bird families, such as kingfishers (*Alcidinae*) [52], pelicans *(Pelicanidae)* [54] and some shearwaters *(Procellariidae)* [55]. These birds do not plunge as deeply, and shearwaters rely more heavily on propulsion from their wings to forage [55].

3.2.2 *Pursuit Diving*

Auks (*Alcidae*), particularly guillemots (genus *Uria*) are almost equally at home moving in air or in water. These birds swim by flapping in both media, while adapting their wing morphology dynamically to their environment. Flight in air favours large wing areas, while underwater flapping is better achieved with smaller, shorter wings [56] and to maximise their efficiency guillemots will morph their wings to a highly swept, lower aspect ratio configuration when swimming (Fig. 3.2E) [45]. A consequence of adapting smaller wings for flapping underwater is an increased wing loading. A Common Guillemot (*Uria aalge*) weighs 0.9 kg and has a wing loading of around 170 N/m^2. This can compared to 55 N/m^2 for a plunge diving Brown Booby (*Sula leucogaster*) of the same mass, or an expected 100 N/m^2 for a 0.9 kg flying fish. The high wing loading impedes the birds' manoeuverability and takeoff performance, forcing them to flap at much higher rates and fly at high speeds (Fig. 3.4).

Other birds such as cormorants, grebes, ducks and loons dive with folded wings, propelled by their feet. Cormorants (*Phalacrocoracidae*) are some of the most capable foot propelled divers, and are widely studied. Adaptations in these birds' bone and muscle structure make their specific buoyancy around half that of other flying seabirds [57] (Fig. 3.4), and cormorants are able to achieve neutral buoyancy by swimming to depths greater than 50m, at which point the air in their feathers and lungs is sufficiently compressed by water pressure [58]. At depths where they are still positively buoyant, cormorants use an intermittent 'burst and coast' swim pattern to counteract upthrust efficiently while swimming [59]. They do not dive from flight, but will execute a half dive by leaping in an arc from the water surface to achieve some initial momentum.

Buoyancy control is important to diving birds because they are obliged to carry with them a certain volume of air when diving. This is for both insulation and respiration, as well as the need for dry plumage when returning to flight. As mentioned previously, in peguins (*Spheniscidae*) the trapped air compressed beneath feathers can escape during ascent from a dive, which may reduce drag by as much as 70% through the creation of a lubricating air film [50]. This allows the birds to leap onto high vertical ice shelves from the water. However, the presence of feathers is overall likely to constitute a drag penalty in steady submerged swimming [60] for thermal insulation, and the ability to fly efficiently [61]. While most pronounced in cormorants, pursuit diving flying birds have all evolved less waterproof plumage than their non-aquatic counterparts to some degree, and many of these birds must dry their wings on land before attempting long flights [62]. The air retaining features of feathers are a result of their hydrophobic microstructure, which is also of great importance for flight from water, where takeoff power requirements in birds can be increased by around 30% with a 10% mass increase from retained water [29].

3.2.3 Diving Insects

Many insects are able to swim beneath the water, and some have also retained their wings and can still fly in air. Beneath the water, diving beetles (*Dytiscidae*) use their legs to kick, with air trapped beneath their wings for breathing (Fig. 3.2F) [63]. Without the soft, compressible air retaining structures of birds, the trapped air forces them to constantly swim downward to counter buoyancy. Across the several thousand known species of diving beetle, a broad range of morphologies exist. These differences can be related to the relative coverage of their habitat by water, and that water's flow speed. Fast swimming beetles exhibit slender, more streamlined body shapes with fewer protuberances, to reduce turbulence [64].

Dragonflies (*Anisoptera*), arguably one of the natural world's most capable flyers, lay their eggs underwater and will dive directly through the surface, but adults are unable to swim and instead rely on crawling along submerged material [65]. While underwater, breathing air trapped by their wing microstructure allows them to remain submerged for hours at a time, but to leave the water, they struggle to free their wings from surface tension and must spend a period drying before further flight. Dragonfly nymphs are fully aquatic, and larger larvae (approximately 50 mm in length) are able to swim using jet propulsion similar to that of cephalopods. By doing so, the animals can produce pulsed water jet thrust at around 2 Hz, with up to 150 mN amplitude. However, the jetting mechanism is tied to the larval form's underwater breathing apparatus, and is abandoned once the dragonfly metamorphoses into its adult stage [66].

Most insects leave the water by simply climbing out and drying themselves, but several very small species of jumping insect have also adapted to leap from the water's surface. Pigmy mole crickets (around 7 mg mass) are able to leap to around 60 times their own height from the water surface [67], but such small insects are reliant on surface tension to support themselves on the water while static, and do not have any swimming ability.

3.3 Flight in Aquatic Animals

Several species of aquatic animal are able to escape predators by gliding above the water, where refraction obscures their movement from pursuers. Flying aquatic animals are limited in their movement above the water by the constraints their biology imposes on their aerodynamic surfaces, the lack of oxygen, and in some instances the presence of more agile (and hungry) birds. Despite these limits, the strategies of multimodal marine life are fascinating, and have much to offer aquatic aircraft deisigners.

3.3.1 Flying Fish

Over 40 species of marine fish have evolved the ability to perform extended flights above the water's surface. These animals leap from the water's surface and glide on enlarged pectoral fins (Fig. 3.2A), and are capable of spending over 40 s airborne [68]. Beneath the water the oceanic flying fish (family *Exocoetidae*) are conventional swimmers, with their pectoral fins folded against their body. But when deployed in flight, these fins form high performance [69] membranous wings. Wing flexibility also allows the fish's body to rotate relative to its wings, providing further stability [70], particularly during emergence. The wing itself has pointed tips, which will reduce induced drag without increasing the root bending moment [71], which must be supported by muscle tension and has an anaerobic metabolic cost. The wing membrane is supported by many cartilaginous fin rays, with an L-shaped cross section [68]. This structure will enhance wing stiffness to weight ratio, but also allows tessellation on folding. The deployed wing is also given a non-smooth surface by this batten structure, which may also offer aerodynamic performance benefits in the low Reynolds number range [72].

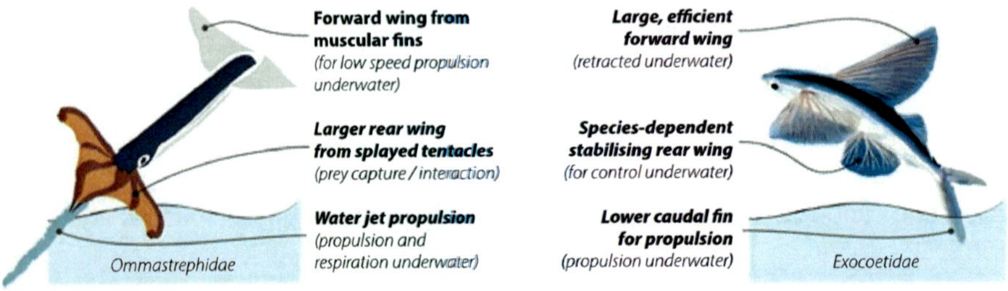

Fig. 3.4 Adaptations for flight in aquatic animals

Underwater, a typical 30 cm flying fish cruises at low speed (1 m/s), but is capable of burst swimming at up to 10 m/s when leaving the water. However, gliding speeds are in the range 15–20 m/s and the fish are better able to reach these high speeds in air, where drag is lessened [68, 73]. Additional speed is then gained through a 'taxiing' phase, with wings deployed clear of the water to balance body weight with lift, while only the caudal fin remains submerged, and can continue to propel the animal (Fig. 3.2A). This also enables prolonging of flights by intermittently taxiing and 'topping up' airspeed, and allows the fish to take more advantage of the ground effect [30]. As is common with most fish, their skin secretes a mucus which is viscosity-reducing, a surfactant and hydrophobic [74]; features which can reduce drag, ease wing unfolding [68] and help to shed water in flight.

3.3.2 Flying Squid

The only fully aquatic animals displaying what can be considered powered flight are the flying squid (several species mainly from the family *Ommastrephidae*, Fig. 3.2C). Flying squid have two sets of deployable wings. Their rear, main lifting surface is formed by their tentacles, splayed apart to stretch a membrane, and the forward wing is formed by a pair of fins, which are also used for flapping propulsion underwater at low speeds [75]. The forward fins of flying squid species have a higher aspect ratio than those of non-flying species, which is likely a result of adapting for flight [73].

To power their flight, these animals use a pressurised water jet to provide thrust in the air [76]. This airborne jet provides the fish with a final momentum boost, allowing them to perform an aerial jump-glide in a similar fashion to flying fish. These flights can last for 7–8 s (a full power jet expulsion by a similarly sized quid lasts around 200 ms [77]) at peak speeds of up to 11 m/s [78]. When swimming underwater, jet propulsion is fundamentally less efficient than locomotion through the beating of fins and tails seen in teleost fish and cetaceans, due to the larger mass of water that must be moved for propulsion. However, the thrust response of jet propulsion is much faster, which gives squid advantages in escape and manoeuvreability [79]. Furthermore, the jetting mechanism is uniquely applicable to aerial-aquatic locomotion because it can produce thrust in both water and air.

The squid's water jet is driven by contractions of its mantle cavity, which can undergo volume changes of 400% during a full power jet cycle. Driving the jet with radial contractions maximises the volume of water a squid can expel relative to its size, and reduces the internal flow velocity and associated viscous losses (relative to a piston driven expulsion, for example). During the jetting phase, the water pressure within the squid mantle reaches around 25 kPa in a typical squid with a 21 cm mantle length [77]. The squid's cavity muscle cannot provide sufficient power over its full contraction range, and so additional elastic energy is stored in collagen fibres as the cavity is expanded and filled with water [80]. The outflow nozzle of the jet is also muscular, and the squid can adapt its geometry during jetting to optimise propulsive efficiency. This also allows it to vector its jet thrust to steer.

As muscular hydrostats, the wing muscle tension required during flying may limit the duration of squid flights [78], and the animals are observed to actively brake and fold their wings to return to the water [81], diving in a streamlined, nose forward configuration. The squid's main wing is formed by soft tentacles, allowing the squid to adapt its planform shape. It is suggested that the squid has 'chosen' a near elliptical planform [73], which would minimise induced drag, but shape adjustments could reduce the wing bending moment [71], suggesting that muscle tension may not be limiting in flight. However, aspects of the planform shape could also be a structural artefact of stretching the wing membrane to maximise wing area.

3.3.3 Sizing and Energetics

In nature, aerial and aquatic animals show marked differences in metabolic costs (Fig. 3.5A). Transport costs in water are reduced because self-weight does not need to be counteracted during motion due to the presence of buoyancy [56]. Figure 3.5A also highlights the importance of size as a design consideration, and as size is reduced increased effects from viscosity, surface tension and natural perturbantions act to reduce locomotion efficiency. However, many of the envisaged applications for an aerial-aquatic vehicle (Sect. 2.2) require a vehicle that can be rapidly transported to an area of interest for deployment, and this will be facillitated by small size and low weight. Smaller robots can also functionally locomote in smaller spaces, which will allow closer observation.

Flight favors low density, while swimming favours neutral buoyancy, and pursuit diving birds lose flying ability as their size increases [86]. The largest auk and cormorant (*Pinguinus impennis* and *Phalacrocorax perspicillatus*) were flightless and near flightless respectively. Within a given locomotion strategy, most aquatic animals show little variation in size and speed (Fig. 3.5B). However, squid range in length from 20 mm to well over 10 m, and flight has been documented in squid ranging from 60 mm to over 1 m [81], a variation in size far surpassing that of any other

Table 3.1 Key design principles from nature

Animal	Dry flight	Water entry	Submerged movement	Water exit	Wet flight
Flying fish (*Exocoetida5e*)	Flight in ground effect to reduce drag [30]. Thin **flexible wings** for stability [69, 72]	**Wings lie flat** against the body [68]	Oscillating tail **fin propulsion,** swim bladder allows changes in buoyancy [56]	Taxiing acceleration with only **propulsive fin** submerged [68]	Hydrophobic mucus helps shed water. Intermittent taxiing to increase speed
Flying Squid (*Ommastrephida*e)	Water jet propulsion provides thrust [76]. Elliptical wing to reduce wing tension [73]	Fore wings lie **flat against the** body. Hind wings streamlined backward [76]	Water jet propulsion [75] **or flapping wings/fins** [75]. Intermittent locomotion used to conserve energy [77]	High speed leap followed by water jet thrust to accelerate [78]	Flying time limited by muscle tension and oxygen, **flight is ended** deliberately by air braking and diving [81]
Gannet (plunge diving) (*Sulidae*)	High aspect ratio wings for **efficient** loitering [82]. Feathered wings adapt to wind perturbations [72]	Wings swept to 90°, air sacs cushion impact [47]. Flight into wind for low groundspeed and fast dive response [82]	Little swimming [52], depth gained by dive momentum, buoyancy used for ascent [48]	Hydroplaning taxiing by **flapping and** foot propulsion	Water repellent feathers shed moisture. Foraging upwind of nest so tailwind aids **return flight,** when food and water add to weight
Cormorant (foot propelled) (*Phalacrocoracidae*)	Relatively high weight compensated for by large wings [83]. Feathered wings	Slowed descent and soft, **floating landing.** Short dive-initiating leap from the surface	Foot propulsion with wings folded. Depth compresses air to reduce buoyancy [58]. Intermittent kicking/gliding **for efficiency** [59]	Hydroplaning taxiing by **flapping and** foot propulsion	Drying period on land [84]. Foraging upwind of nest **for efficient return flight**
Guillemot (wing propelled) (*Alcidae*)	Small wings adapted for swimming, [45] compensated for **by fast flapping and short flights** [82]. Feathered wings	Slowed descent and soft, **floating landing.** Dive begins from standstill	Flapping propulsion, with wings morphed shorter [45]. Depth compresses air to reduce buoyancy	Hydroplaning taxiing by **flapping and** foot propulsion	Drying period on land [29]. Foraging upwind of nest **for efficient return flight**
Diving Beetle (*Dytiscidae*)	**Insect flight, by rapid flapping of** hind wings	Dive begins from standstill. Wings fold beneath forewings/shell [85]	Kicking foot propulsion with wings folded. Air trapped beneath wings for breathing	Flight from standstill on land	Drying period on land

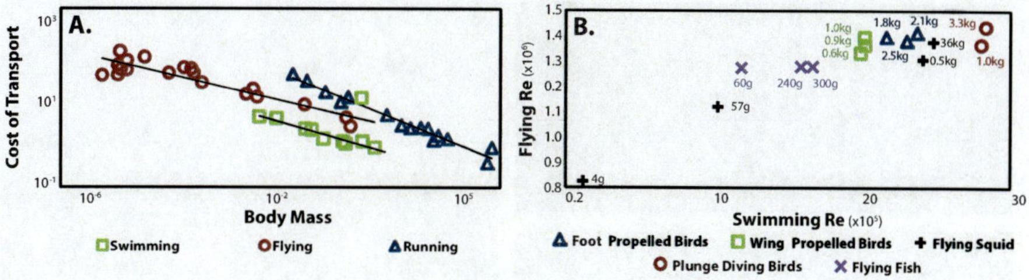

Fig. 3.5 (**A**) The metabolic costs of different locomotion strategies (reproduced from [56]). (**B**) Average airborne and submerged Reynolds numbers for aerial-aquatic animals, with body mass indicated

aerial-aquatic animal. This suggests a scalability that could make taking inspiration from flying squid highly applicable to robot design.

It has also been suggested that flight in squid and fish offers energetic advantages for migration over aquatic locomotion alone [76]. This is due to the reduced drag in air, and is against the overall trend in Fig. 3.5A. While the subject of contention, this highlights the fact that size is not the only consideration for transport efficiency, and careful selection of locomotion strategy is key to the design of a multimodal vehicle.

3.4 Design Insights from Nature

Flying fish and birds are able to leap directly from the water, avoiding waves, obstacles and taxiing takeoff complications. However, flying fish must first reach submerged speeds of around 20 body lengths per second. This performance represents an order of magnitude increase over what has been achieved up to this point by biomimetic underwater robots [87, 88], and the power availability of small scale mechanical actuation systems cannot yet match that of bird and fish muscle, without prohibitive sacrifices in frequency response or stroke rate. This will make it hugely difficult to takeoff from the surface of the water by flapping, or biomimetic swimming (Fig. 3.6).

Jet propelled launch is uniquely applicable to short take-offs by aquatic aircraft. Compared to flapping and beating, a jet is mechanically simple, and jets of mass have a very rapid thrust response making them ideal for an impulsive takeoff. Importantly, jets can continue to produce thrust when clear of the water surface and its associated high drag, unlike fin based water surface jumps, which rely on reaction forces from the surrounding fluid. This allows a vehicle to escape the water and accelerate when airborne, where drag is dramatically reduced compared to in water, and pass directly through surface waves which could prevent other types of takeoff. This allows an airborne transition stage without the need for the surface taxiing phase required by birds and fish, and when on water, requiring far less takeoff area than taxiing takeoffs. While this could also be achieved with combustible rockets, rocket fuels are highly controlled by regulations, are difficult to miniaturise [89] and many situations (an oil

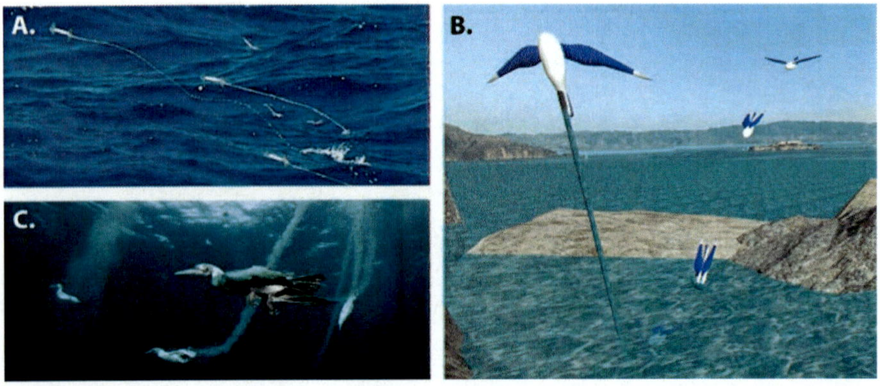

Fig. 3.6 Bioinspired principles for AquaMAV: Plunge diving by gannets (**A**), Jet propelled take off by squid (**B**), Plunge diving AquaMAV concept sketch. *Images used with permission from: Alexander Safonov* (**A**) *and Bob and Deb Hulse* (**B**)

rig accident, for example) would preclude the use of uncontained fire by exploring robots. A water jet potentially offers a clean and safe alternative.

3.4.1 Wing Folding

To the best of the authors' knowledge, no truly fixed-wing animals exist, and wing folding is a feature of most flying creatures. But it becomes yet more important when locomotion in water is required. Folding wings reduce a vehicle's profile underwater, which can greatly reduce drag. Folded wings also do not have to support as large a hydrodynamic load, which reduces structural requirements, and hence mass. Of the animals discussed in this chapter, only the guillemot does not fully collapse its wings when in the water, and these birds must still morph their wings into a much lower aspect ratio configuration in order to swim. It is reasonable to then suggest that for a winged aerial vehicle to submerge, some degree of wing folding will be obligatory.

In robotics, rigid winged underwater gliders do exist, but these craft are too heavy, with wings too small for aerial flight without very powerful propulsion. Many existing aerial vehicles do employ some kind of morphing wing [90, 91], and others feature deployable wings [92]. However, wing deployment systems are often merely for storage, without the ability to refold, and most morphing concepts only vary the wing geometry slightly. Implementing high performance wing folding into an aerial-aquatic vehicle will likely require improvements in mechanical design over existing systems. It is additionally difficult to implement a wing fold that gives a large planform change while also accommodating control surfaces within the wing.

Flying fish wings are a mechanically attractive inspiration source for deployable wing design. Their wing structure, with rigid battens supporting a flexible membrane is reminiscent of recent MAV designs such as [23, 93]. The wings are also attractive when considering deployability, because they are actuated entirely at their root and do not contain any internal musculature, unlike the wings of birds and bats which are significantly more complex than what can be achieved with current actuation and sensing technology.

3.4.2 Plunge Diving

Among aquatic birds, plunge diving birds show the least detriment to their flying ability as a result of adapting for the water, and their landing strategy is robust, simple and does not require additional propulsion or neutral buoyancy to penetrate beneath the water's surface. Especially with regards to mechanical design feasibility and control complexity, the gannet's plunge dive represents a robust and practical strategy for entry, just as the squid's jet propulsion system is an effective means of takeoff.

The requirements of accurate control and a large area during taxiing takeoff also exist for a soft water landing. However, plunge diving in the manner of the gannet avoids the complications of soft landings. A plunging vehicle would need minimal landing area, and would need only a GPS waypoint to execute a landing in an area of known bathymetry. Dive velocity can be controlled by varying initial dive height, and will allow accommodation for shallow water.

In order to design for effective plunging, folding wings will again be essential, and the gannet's strategy of sweeping wings fully backward at entry is a mechanically simple way to implement this. Such a wing folding system should ideally be designed such that folding the wings shifts the centre of buoyancy aft of the centre of mass. This would that the dive would be stable under buoyancy forces, but at the termination of the dive, unfolding the wings would move the centre of buoyancy forward and orient the vehicle nose up for launch.

3.5 Conclusion

Nature provides several examples of how the biomechanical conflicts of swimming and flight may be resolved. However, biological necessity imposes constraints that do not translate to robot missions. Equally, many attributes of natural systems that enable locomotion (e.g. controllable muscle stiffness, distributed mechanical sensing, highly hierarchical materials) are difficult to replicate in robots. Nature can only provide insight - it does not prescribe any design, and judicious engineering must also account for the needs and capabilities of modern technology.

Chapter 4
Synthetic Aerial Aquatic Locomotion

4.1 Multimodal Robots

A wealth of research exists into the broader question of how robotic mobility can be expanded beyond a single domain/terrain. A significant amount of recent research attention has been given to the implementation of aerial-terrestrial mobility into miniature robots [94], resulting in mobile robots with shared subsystems and additional mechanisms which are analogous to the AquaMAV robot presented in Chap. 7. Robots such as MMALV [95] and DALER [96] have integrated additional appendages into fixed wing flying vehicles, allowing them to land and move on ground. As with aquatic locomotion, the terrestrial movement speed is far lower than that required for flight, and as a consequence, MMALV is not able to transition to flight from ground unassisted, while the morphology of DALER precludes flight while walking and the robot is unable to take off, though hover capability is intended for a future prototype.

Once of the most inmediate consequences of the implementation of multiple modes of locomotion into a mobile robot is the increase in mass resulting from the additional subsystems required. In [9], a performance metric is proposed based on the weight of components unused during one mode of operation. This is effectively a measure of weight reduction, and is of particular consequence for flying robots, for which weight strongly influences the lifting area and power requirements of flight. For underwater locomotion, buoyancy is more inhibiting, so the relative density and volume of the robot has a stronger effect, but when taking off from water, inertia dominates and extraneous mass will increase the power requirements of takeoff (this is exemplified by the penalty to jumping performance caused by the addition of gliding to a powerful miniature jumping robot in [97]).

The extra subsystems of a multimodal robot can have benefits when inactive, in spite of the additional weight. In the crawling and flying robot, 'Bolt' [98], the presences of wings stabilises the crawling mode of locomotion, and helps the robot to achieve the ground speed necessary for take off. This is explored further in [99], in the context of the evolution of flight in terrestrial animals. Similarly, the provision of

R. Zufferey et al., *Between Sea and Sky: Aerial Aquatic Locomotion in Miniature Robots*, Biosystems & Biorobotics 29, https://doi.org/10.1007/978-3-030-89575-4_4

wings can help jumping robots to land in the correct orientation and prevent landing damage by slowing descent speeds [9, 100, 101].

Producing an aerial-aquatic system that is capable of active propulsion in both air and water is highly challenging, with only a few successful prototypes presented to date. The remainder of this chapter presents a concise overview of the main types of aerial-aquatic robots developed so far, and discusses the key features of each type.

4.2 Aerial Water Sampling

Using larger multirotor platforms, several projects have demonstrated the efficacy of water sampling with UAVs, using a suction pump hung around 1 m below the vehicle for water sample ingestion [102, 103], as shown in Fig. 4.1. This approach relies on accurate sensing and control to maintain position in the presence of ground turbulence while the sample pump is lowered. However, at the small scale, a fixed wing vehicle provides greater range and speed than hovering vehicles. For example a plunge diving approach would reduce the need for accurate control, which allows for platforms to be produced at lower cost and operated in larger numbers.

While not multimodal, systems also exist for deployment of underwater vehicles from the air. Such systems are reliant on single use, disposable equipment such as parachutes, airbags and discardable wings (Fig. 4.2C). However, the existence

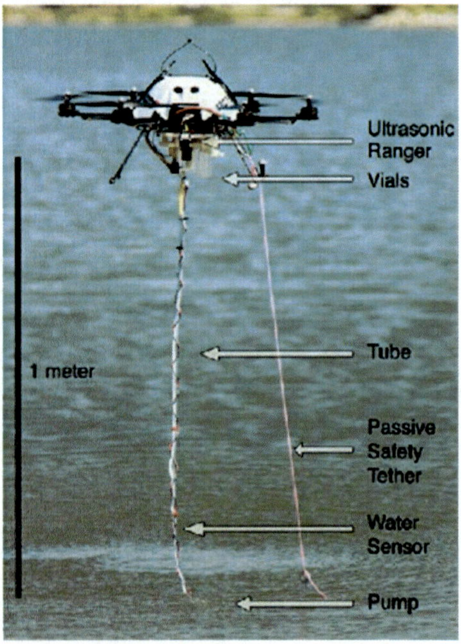

Fig. 4.1 Autonomous Aerial Water Sampler, from [102]

Fig. 4.2 Aerial-aquatic robotics. Aquatic unmanned aerial vehicles: (**A**) The University of Michigan's 'Flying Fish' and (**B**) WarriorAero's 'Gull'. (**C**) TACMAV underwater MAV launcher. (**D**) 'Sea Robin' submarine launched UAV launch sequence. (**E**) Dropsonde air launched atmospheric reconnaissance device.*Images used with permission from: University of Michigan* (**A**), *Warrior AeroMarine* (**B**), *US NCAR* (**C**), *NAVSEA-AUTEC* (**D**), *ARA Force Protection* (**E**)

of a demand for air launched underwater vehicles [18] is important nonetheless, and Aerial-Aquatic MAVs would have inherent capability for air launched aquatic response.

4.3 Seaplanes

Many robotic aerial vehicles exist which use the water's surface as an extended runway. Three examples of UAV seaplanes are the Oregon Ironworks 'SeaScout', Warrior Aeromarine's 'Gull' and the University of Michigan's 'Flying Fish' (Fig. 4.2A/B). The former two vehicles are remotely operated, but are able to land and takeoff autonomously, while the Flying Fish is fully autonomous and is designed to operate as a mobile surveillance buoy [104]. All of these aircraft are propeller driven, use a conventional wing layout, execute soft landings and are incapable of fully submersing themselves in the water. They are also well out of the micro scale, all having wingspans of around 2 m. To takeoff, these vehicles rely on highly buoyant pontoons to keep their propellers positioned clear of the water surface, require large landing and takeoff areas and are not able to move through difficult terrain, such as flooded buildings or areas of floating debris.

Several military programmes have attempted to create manned submersible aircraft [105, 106], but to the author's knowledge these endeavours have not been completed. In the past decade, the recent shift towards unmanned technology has opened up opportunities for MAV designers. The US Navy Research lab has recently demonstrated the successful launch of a UAV from a submarine missile tube (Fig. 4.2D). The vehicle (called 'Sea Robin' UAS) uses a discardable tube which brings the vehicle to the surface and provides launching power. At a smaller scale, the ARA Inc. 'TACMAV' can also be deployed from below the surface, by divers using an airbag and compressed gas to launch an MAV (Fig. 4.2E). Both of these vehicles feature deployable wings, but neither have any true aquatic capability and are solely aerial vehicles after takeoff. However, these projects do demonstrate the efficacy of an impulsive leap into flight from the water, if sufficient power can be provided on-board.

The Gull UAV (Fig. 4.2A) does have the ability to fold its wings, though this is used for manoeuvring and storage rather than diving. It also is able to exploit surface waves (0.3 m, or 15% of its size) at takeoff, using them as ramps.

4.4 Rotary-Wing Vehicles

Multirotors can provide a robust option for undertaking aerial-aquatic locomotion, thanks to their simple structure and lack of actuation surfaces. Due to their vertical take-off and landing capability, they can land on water and sink, without requiring added design features. Several multirotors have achieved a degree of aerial-aquatic mobility. Although various systems have demonstrated the ability of submersible locomotion, the use of aerial propulsion system in off-design conditions results in the system operating at very low efficiencies (see Chap. 10). More broadly, rotary wing vehicles are an inefficient form of locomotion, and must use power continuously to stay aloft, which is why considerable research effort has been invested in perching systems [107]. Aquatic locomotion is helpful here, as bodies of water provide widely available, large landing surfaces. However, a drawback is that these vehicles require a calm surface to take-off, while their principle advantage is that they allow multiple swimming and flying phases in a single mission.

Table 4.1 shows a comparison of the main solutions presented up until 2020 in peer-reviewed publications. Figure 4.3 [108] consists of a 2 kg drone with two layers of 4 propellers, all identical. The double-layer layout theoretically guarantees that at least half of the rotors provide thrust during the transition. With this technology, no velocity is required to clear the propellers from the water, permitting a slow, sequenced transition. This research finds that 8 identical propellers can be used, albeit at the expense of underwater efficiency, due to the low rotational speeds required in water. The researchers demonstrate a full prototype that performs aerial and aquatic locomotion with a tether. The use of a tether suggests limitation in payload of the complete system, perhaps due to the addition of waterproofing structures. The presented prototype paved the way towards employing multirotors in air and in water.

Table 4.1 Operation envelope of prototypes presented in this book

	Naviator [108]	Looncopter [109]	HUAUV [110]	NEZHA [111]
Weight	2 kg	2.7 kg	2 kg	4 kg
Structure	Quadcopter with water propellers	Quadcopter	Quadcopter with water propellers	Quadcopter with fixed-wing
Actuation	8x brushless	4x brushless + piston ballast	8x brushless	4x brushless + compressed gas bladder
Air	8 propeller	Propeller	Propeller	Propeller + wings
Water	8 propellers	Propeller + ballast	Water propeller	Gliding
Transition	Double set of propeller	Propeller + floating	Double set of propellers	Propeller + floating
Propeller	NA	30 cm	60 cm diameter	36 cm
Demonstrated operation	Tethered	Full 15 min cycle	Simulation only	Individual stages shown

Fig. 4.3 (**Left**) Naviator2 robot. (**Right**) Looncopter robot

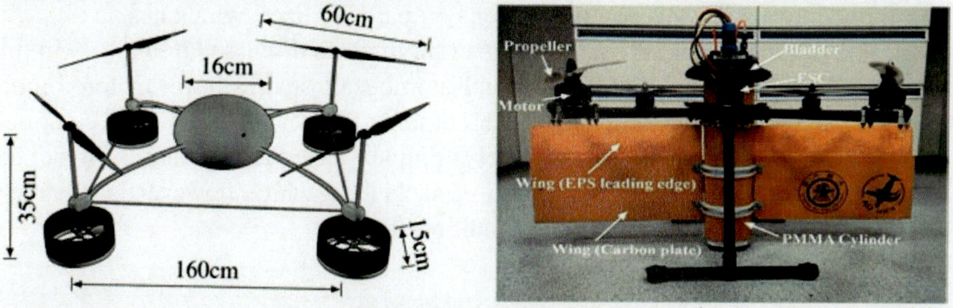

Fig. 4.4 (**Left**) HUAUV Concept. (**Right**) NEZHA robot

In a similar structure, the Fig. 4.3 [109] replaces the double row of propellers in favour of a buoyancy control mechanism. With adjustable buoyancy, the multirotor can float sufficiently to clear the four propellers from the water surface during transition back to flight. The innovative buoyancy control system maintains authority over the tilt angle of the whole robot. The system is designed so that the robot floats in an upright position, while underwater it will take on a 90° orientation for horizontal movement.

Both the Naviator and the Looncopter robots present effective and practical solutions that have been demonstrated to work experimentally. However, the efficiency of using an air propeller in water is low (see Chap. 10). This limits the usability of such robots to short underwater missions. The Fig 4.4 [110] intends to overcome this limitation by replacing the lower row of propellers with aquatic propellers, solely intended for underwater motion. The resulting concept is presented in simulations and the control strategy is detailed in further publications. No working demonstrator has been shown to date.

Lastly, a team of researchers has presented a Vertical Take-off and Landing (VTOL) configuration [111], increasing aerial efficiency while still exploiting some of the advantages of multirotors. This hybrid flight concept is capable of vertical motion, particularly important for the transition to and from the water, as well as of efficient fixed-wing flight. The Fig 4.4 additionally features a compressed gas system that adjusts the buoyancy when underwater. Interestingly, the underwater locomotion does not use the propellers, which would be inefficient in water. Instead, the wing is used to glide up and down at ±45°, with the vehicle driven solely by its buoyancy system.

All of the solutions presented in this section rely on clearing the vehicles' four propellers from the water to transition to the air. While the specific methods differ, the clearance achieved is in the order of centimetres. This limits the applicability of such solutions to calm waters (waves below 10 cm). While underwater propulsion can be achieved using the same propellers as for flight, this comes at a performance cost and does not always lead to effective locomotion. By contrast, the use of separate propellers for the aerial and aquatic mission phases implies a payload penalty, as unnecessary additional weight needs to be carried. Lastly, multirotor vehicles remain constrained by a short mission time when compared to fixed-wing ones.

Despite these drawbacks, the benefits of multirotors discussed previously in this section are still considerable, and a small number of studies have explored more unconventional solutions that attempt to combine the best of different locomotion approaches without becoming excessively complex. One such solution is in fact the hybrid fixed/rotating-wing robot by Lu et al. [111]—which however is relatively complex compared to an off-the-shelf multirotor.

4.5 Fixed-Wing

Just like multirotors can be used for aquatic operations, aerial-aquatic operations with standard fixed-wing configurations have been demonstrated. Compared to multirotors, fixed-wing vehicles have the advantage of a better endurance and long-distance flight efficiency, which would potentially allow for longer-duration missions. However, using fixed-wing vehicles underwater requires more extensive adaptations compared to the multi-rotor case. One issue, as discussed in Chap. 1, is buoyancy. While a thicker, profiled wing may be useful for flight, it will make achieving neutral buoyancy challenging. Fixed-wing vehicles are also generally more difficult to manoeuvre underwater. Lastly, transitioning out of the water constitutes a particular challenge for such vehicles, requiring either strong propellers and elaborate control, for a gradual water exit manoeuvre, or the rapid release of a large amount of energy, for rapid, impulsive water exit.

Nonetheless, a number of fixed-wing aerial-aquatic robots have been developed. Moore et al. [112] developed a flying-wing MAV capable of controlled, gradual water-to-air transition. Largely based on off-the-shelf components, and relying on a standard propulsion system, the MAV exits the water via a prop-hanging manoeuvre. A closed-loop control scheme was developed to allow the vehicle to track an optimal water exit trajectory. While effective, the water exit scheme requires a calm water surface, and the elaborate control scheme relies on a complex model that needs to be highly accurate. Small deviations from the required initial condition of the vehicle were shown to lead to failure. To avoid the structural challenges of water impact on wing structures, the robot adopts a low-aspect ratio delta configuration (as in Chap. 9), which has intrinsic structural strength and lacks the abrupt cross-sectional area change found in a rectangular wing. The benefits of a delta wing come at the expense of aerodynamic efficiency.

Stewart et al. [113] have also presented a fixed-wing aerial-aquatic robot, this time with an aerodynamically efficient rectangular wing, with a thick airfoil. The wing is allowed to fill with water to counter buoyancy, and the propulsion motor has sufficient power to lift the aircraft vertically out of the water and into the air, with the water draining passively from the wing. The vehicle's buoyancy is trimmed such that it passively reorients itself nose-up for take off. To allow underwater locomotion, the vehicle relies on hydrodynamic lift to oppose buoyancy. As a consequence, the UAV cannot stop moving beneath the water, and relies on flight momentum to generate initial depth. Once the vehicle is stationary in the water, a short aerial phase is required to move beneath the surface.

Recognising the structural challenge of impacting the water with high aspect ratio wings, studies have recently been conducted on the implementation of plunge diving capability into a flying vehicle [25, 114, 115]. Liang et al. [25] have developed a testing platform able to fold its wings before diving into the water. However, this device is currently an experimental platform being used to evaluate the requirements for a plunge diving vehicle and does not yet have powered flight capability.

Folding wings for plunge-diving have now been implemented on two aerial aquatic platforms, the AquaMAV described in this book (Chap. 7 [6]), and the more recent 'Dipper' UAV [116], which combines the propulsion and plunge diving concepts in Chaps. 10 [8] and 7, but with separate propellers for air and water, driven by a single motor with sufficient thrust for escape without requiring jet propulsion, and an integrated active control approach for transition from air to water, rather than relying on a passively stable impulsive transition (e.g Sect. 6.8).

4.6 Tilting Propeller configurations

Peloquin et al. [117, 118] avoid the need for taxiing space by allowing the vehicle's propeller to rotate 90°. In this way their flying wing aircraft can swing itself out of the water and accelerate into the air, with the propeller passively rotating back to level flight. However, this launch process is only successful over a narrow condition, and is likely to struggle in rough surface conditions. The vehicle is highly buoyant and cannot submerge, but does dive directly into the water to avoid the complications of a soft landing, and can use its rotatable propeller to self-right afterwards. This innovative solution elegantly lands and takes off from the water with few actuators, an advantage to improve reliability, especially in aquatic environments.

4.7 Flapping Wing Vehicles

Flapping propulsion has been demonstrated in many robots, and has many advantages, both in air [119] and in water [120]. Lock et al. [45] used experimental data to develop a model for the effect of wing sizing and morphing on the power requirements in air and water of a guillemot inspired flapping wing robot. This enabled them to propose different optimised geometries depending on the relative amount of time the robot would spend in air and water during a mission, but this work was not developed into a practical robot. Izraelevitz et al. [121] also investigated the use of flapping foils for dual mode propulsion, using changes in kinematics rather than morphology to ensure efficient thrust production in different media, and demonstrated the efficacy of this strategy in tunnel tests.

On a much smaller platform, [122] demonstrated aerial-aquatic locomotion with an insect scale vehicle. Flapping wings are used for both aerial and aquatic propulsion, with changes in stroke rate resulting in dynamically similar flows in both media. However, the robot was powered externally so energetic efficiency was not of immediate concern, and the robot was only able to operate in de-ionised water. The robot was initially unable to transition in or out of water; a surfactant was used to allow the robot to submerge, and the robot could not subsequently leave the water. However, a further developed version of the aquatic Robobee system demonstrated aquatic escape using an electrolysis chamber to separate water into hydrogen and air that could be ignited to allow transition, by providing the impulse needed to break free of the water surface and begin flight.

While this work has shown that aerial and aquatic flapping propulsion is feasible with a single system, the ability to transition between modes remains elusive. For a flapping vehicle, the loss of stroke range when floating on the surface would make it highly difficult to take off without pontoons, the use of which would then hugely inhibit submerged movement. With current technology, hybrid flapping propulsion is not likely an immediately feasible design strategy for an aerial-aquatic robot, principally because of the difficulty in making such a system transition between locomotion modes. However, future advances may make this more feasible.

4.8 Possible Transition Methods Between Water and Air

Multi-modal aerial-aquatic robots all demonstrate transitioning from water to air. Water to air transition methods can be classified into three broad categories. Firstly, *impulsive* water escape requires storing a certain amount of energy that can then be released extremely quickly. Secondly, horizontal *lift-off* relies on aerodynamic surfaces to generate lift (wings, propellers). Thirdly, *vertical take-off* requires strong propellers to lift the vehicle vertically out of the water. Somewhat analogously, air to water transition approaches also typically fall into three categories, i.e. (i) rapid *diving*, *horizontal landing* on the water surface, and (iii) *vertical landing* and sinking into the water. The advantages and limitations of each method are summed up in Tables 4.2 and 4.3, and will be explored further in the subsequent chapters of this book.

Table 4.2 Aerial-aquatic water-to-air transition (WAT)

	Impulsive	Lift-off	Vertical take-off
Principle	Fast and large energy release from chemical or elastic element	Aerodynamic surfaces create lift	Propellers create vertical thrust
Cost of method	Additional on-board power storage or production	Take-off distance required	Uncertain transition or floats
Advantages	No control required, insensitive to sea-state	Efficient maneuver	Mechanically simple

Table 4.3 Aerial-aquatic air-to-water transition (AWT)

	Diving	Landing	Vertical landing
Principle	High-speed, head-first dive	Controlled, horizontal landing	Propellers create vertical thrust
Cost of method	High structural requirements	High level of control	High power requirement
Advantages	No control required, insensitive to sea-state	Does not require submerging, low forces	Gradual approach, mechanically simple

Chapter 5
The Physics of Aerial Aquatic Locomotion

This chapter presents an overview of some fundamental physical laws and concepts at play in generic, as clarified in Figs. 5.1 and 5.2. The vehicle-specific physics are then introduced in the following chapters and form the basis for locomotion derived for the different vehicles presented.

5.1 Resolving Contradictory Design Pressures

Aerial-aquatic locomotion requires to solve in one system both the constraints for flight and the constraints for aquatic movement.

The Reynolds number is a particularly important metric in the study of aerial-aquatic vehicles. The impact of the fluid on a moving object can be compared at the same Re. A rigid robot's aerodynamic surface geometry is optimal for a given Re number. The Reynolds number is defined as:

$$Re = \frac{\rho_F u L}{\mu} \tag{5.1}$$

where ρ_F is the density of the fluid, u is the flow speed, L is the characteristic length and μ is the dynamic viscosity of the fluid. In an intuitive sense, the Re characterises the relation between inertial and viscous forces.

When a fully rigid moving object transitions from air to water, the density ρ_F increase by 1000 and the dynamic viscosity μ increases by 50, leading to an increase of Re of 20. This difference implies that adaptation is required in order to continue to efficiently move after media transition, either in velocity or in geometry.

R. Zufferey et al., *Between Sea and Sky: Aerial Aquatic Locomotion in Miniature Robots*, Biosystems & Biorobotics 29, https://doi.org/10.1007/978-3-030-89575-4_5

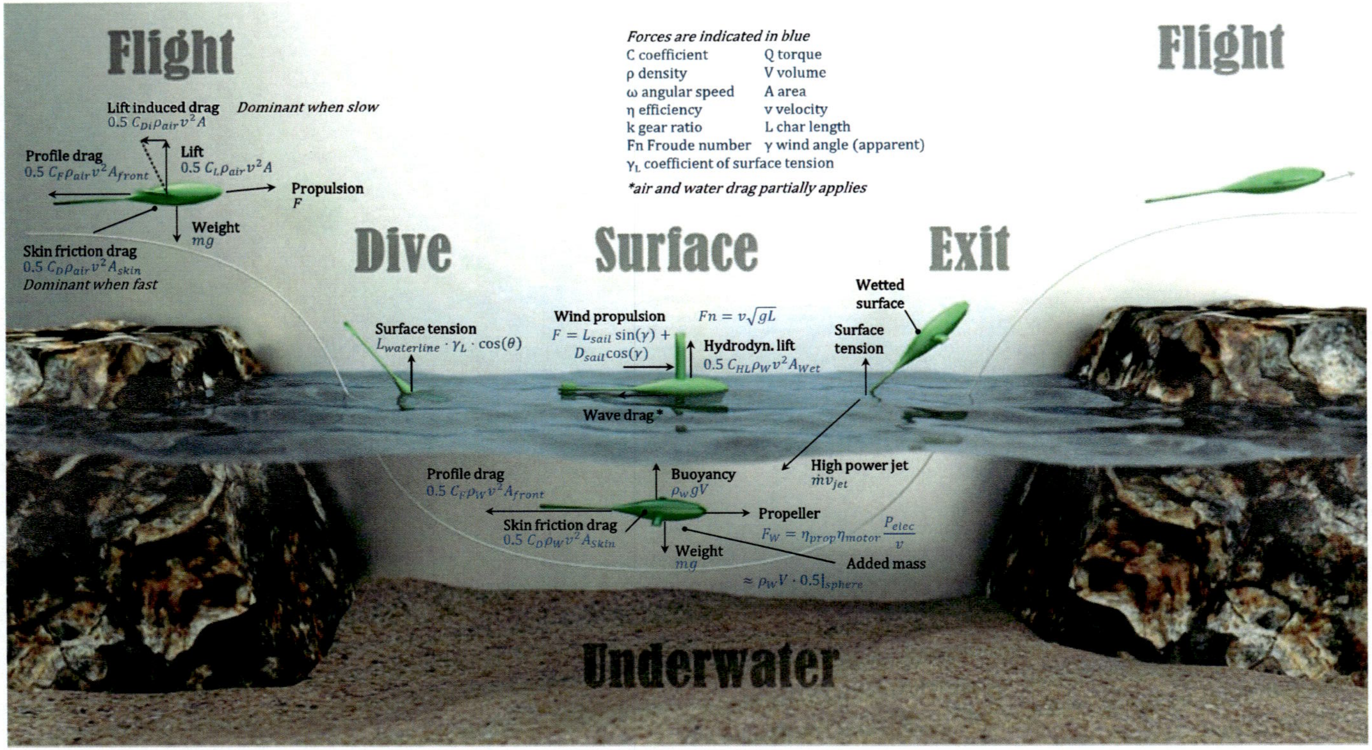

Fig. 5.1 Overview of the physics at play at various points in the aerial-aquatic cycle, specifically the flight, underwater and surface conditions as well as the transitions in-between

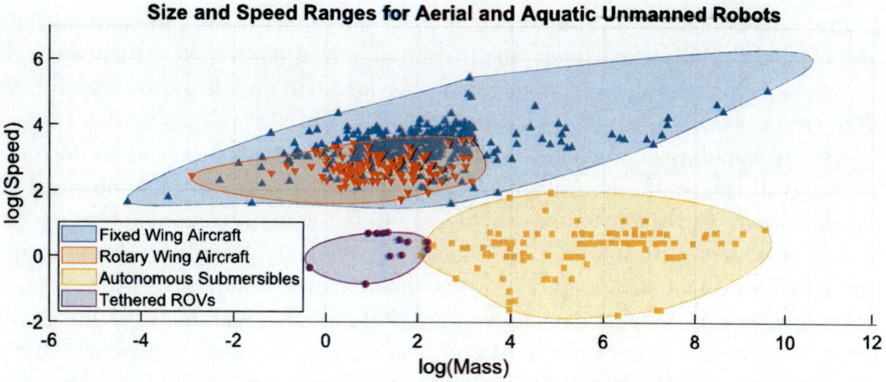

Fig. 5.2 The size and speed of aerial and aquatic vehicles. Aquatic vehicles do not need to support their own weight but must sustain large external pressures, which means that they tend to be dramatically heavier. Aerial vehicles need to generate lift, which requires large areas, light weights, and speed. In the figure above, there is no overlap between aerial and aquatic vehicles

5.2 Aerodynamics

During the aerial phase of their operation, aerial-aquatic robots in many ways behave analogously to standard UAVs, with a number of further particularities due to their additional aquatic locomotion capability. This section provides a basic overview of the aerodynamics concepts most relevant to aerial-aquatic robots, to the extent required for an understanding of the second part of this book. For in-depth aerodynamics theory, we refer the interested reader to one of the numerous aerodynamics textbooks available in the literature (e.g. [93, 123, 124]).

The most common types of small-scale UAVs fall into either the fixed-wing or the multirotor category. In fixed-wing UAVs, lift and thrust forces are essentially separate, with the wings generating the lift force to keep the vehicle aloft and propellers providing the thrust for forward motion. In multirotor UAVs, on the other hand, all forces come from the propellers. Note that the propeller blades are essentially rotating wings, and therefore also produce aerodynamic forces.

Aerodynamic forces are generated when a body moves through a fluid and are conventionally resolved into a lift force component, perpendicular to the velocity of the fluid, and a drag force component parallel to the flow and acting against the movement of the body. Aerodynamic forces are mainly dependent on the shape of the body, its orientation with respect to the fluid (given by the angle of attack, α), the velocity of the fluid with respect to the body (freestream velocity, V_∞), and the density of the fluid (ρ). This dependence can be written as,

$$L = \frac{1}{2}\rho V_\infty^2 S C_L \tag{5.2}$$

$$D = \frac{1}{2}\rho V_\infty^2 S C_D \tag{5.3}$$

In the above equations, S is the wetted area, i.e. the surface area exposed to the fluid, and C_L and C_D are, respectively, the lift and drag coefficients. As evident from the equations, a higher velocity and larger surface area will both lead to higher lift and drag forces, as will operating in a denser fluid.

The lift and drag coefficients are dimensionless values that depend on the angle of attack, the shape of the wing cross-section (airfoil), the Reynolds number and the Mach number. The Mach number effect is negligible as aerial-aquatic robots operate at low velocities. Lift and drag coefficients are typically obtained experimentally, and there are established empirical approximations for simple geometries such as flat plates, as well as large databases [125, 126] available with the experimentally-determined coefficients of the most common airfoils. It should however be noted that especially smaller MAVs tend to operate at low Reynolds numbers, for which available data may be more limited, so that experimental testing may still be required in some cases.

The main characteristics of an airfoil are its chord (straight line connecting leading and trailing edges), camber (line connecting leading and trailing edge and equidistant from upper and lower wing surfaces) and thickness (distance between upper and lower surface, typically represented in terms of maximum thickness and relatively to the chord). Increasing airfoil camber or thickness both lead to higher lift as well as higher drag. These two properties are particularly relevant for aerial-aquatic robots, as they inherently influence the volume of the wing, which in turn determines the amount of buoyancy generated during underwater operation. Buoyancy considerations frequently require a trade-off. A profiled wing may for instance be advantageous for flight performance but undesirable underwater. It should also be considered that requirements may vary. Thus for instance, depending on the configuration, neutral buoyancy may be desired to minimise the power needed to keep the vehicle underwater, or positive buoyancy may be desired to assist return to the surface and water exit. Further to this, the lift itself produced by the wing needs to be considered in the context of both aerial and underwater operation, where requirements may be very different (cf. Sect. 5.1).

Multirotor MAVs are kept aloft by multiple propellers, which not only generate lift for sustained flight, but also allow for effective manoeuvring. Propeller blades are effectively rotating wings, which implies that multirotor vehicles too are significantly influenced by the differing fluid dynamics in air and water. The implications of these differences on propeller performance and efficiency are discussed in the subsequent section.

5.3 The Propulsion Problem

Most people reading this book will have an intuitive sense of the differences in shape between a typical ship's propeller and a typical aircraft propeller. But why is this the case? The blade-tip Reynolds number of the 300 W motor (E320) in the popular DJI Phantom UAV is 10^6, which is approximately the same as the blade tip Reynolds number of the widely used T200 underwater vehicle thruster, which also has a 300 W power rating. The difference is not a change in flow condition but the three orders of magnitude difference in density. In this example the aquatic motor is producing approximately ten times the thrust of the aerial motor at the same power, which means ten times the force acting on each blade.

This force increase, combined with the threat of cavitating water flow is what drives the broad, squat blade design found in the archetypal ship's propeller—the need to sustain large forces while limiting propeller speed below the cavitation speed (cavitation speed depends on the ambient water temperature and pressure). However, scaling down to sizes typical of mobile robots greatly alleviates these problems—smaller blades are both intrinsically stronger relative to fluid forces, and have lower tip speeds.

In fact, aerial propellers may be operated in water without any significant loss in fluid dynamic efficiency, and a variety of research has demonstrated aquatic mobility for multirotor aircraft by simply waterproofing the airframe. The principal problem is in matching motor and propeller torque demands across the different media (see Chap. 10).

5.4 Surface Hydrodynamics

This book investigates both aerial-aquatic robots that completely cross the water surface and robots that stay at the surface. Broadly speaking, a vehicle moving at the water surface will see both the effect of aerodynamics and the effect of hydrodynamics, in function of the level of immersion. However, additional physics phenomena occur to a robot located at the interface between both fluids.

At low speeds, the weight of the object is supported by buoyancy and surface tension, see Eq. 5.4.

$$mg = F_{buoy} + F_{ST} \tag{5.4}$$

$$F_{buoy} = \rho_F g V_{immersed} \tag{5.5}$$

$$F_{ST} = 2\gamma L_{float} sin\Theta \tag{5.6}$$

where the vertical buoyancy force F_{buoy} is given by Eq. 5.5 and the surface tension force contribution is determined by Eq. 5.6. The density of the fluid is ρ_F, the gravity g and the mass m. The surface tension is proportional to the water line length L_{float}. The coefficient of surface tension is $\gamma = 72.8$ mN/m for water at ambient temperature. An important factor contributing to the surface tension force is the contact angle Θ. This angle depends on both the material of the object and the fluid. It determines the wettability of the object. A 0° contact angle implies an hydrophilic material while a 180° contact angle implies super-hydrophobicity (in the case of water as a fluid).

It can be influenced with surface coatings, chemical surfactants and additives to the liquid, drastically changing the magnitude of this force.

The mass of the floating object scales with L^3, whereas the water line length L_{float} scales linearly. Therefore, as the size increases, the surface tension contribution quickly becomes negligible. At gram-scale or below, surface-tension plays a dominant role [122]. The robots presented in this book are much larger (above 100 g), hence surface tension can be neglected.

The immersed volume $V_{immersed}$ causes a large drag forces opposite to the direction of motion. The frontal drag is dominant over the skin friction drag. As the speed increases, the hydrodynamic lift exerts a vertical force on the moving object, lowering the immersed volume. Therefore the frontal area reduces and the skin friction drag becomes dominant. However, as the craft speeds up, it generates a wave after the bow. Its wave length compared to object length increases with speed. At higher speeds, the wave length becomes much larger than the length of the craft, which is then said to be in hydroplaning regime. The Froude number Fn, given by Eq. 5.7, relates the speed of a boat to its length. It indicates the wave-making regime in which the object is situated. It is a key number to scale a moving robot whilst keeping the same wave-making regime.

$$Fn = \frac{v}{\sqrt{gL}} \tag{5.7}$$

where the v is the robot speed, g the gravity and L the characteristic length.

In hydroplaning, the velocity of the craft is sufficient for the hydrodynamic lift forces to enable positive trim angles and the emergence of the bow and the majority of the hull from the water. The efficient transition from rest to planing is a complex process which depends on many geometrical and surface parameters but the hydrodynamic lift can be estimated as follow.

The hydrodynamic lift F_{HL} is written as:

$$F_{HL} = \frac{1}{2} C_{HL} \rho_W A_{wet} v^2 \tag{5.8}$$

where ρ_W is the water density and A_{wet} is the wetted area. The coefficient of hydrodynamic lift C_{HL} is estimated by:

$$C_{HL} \approx \frac{\Delta_0}{\rho_W b^3 Fn^2} \tag{5.9}$$

where Δ_0 corresponds to the craft's displacement and b the width. The characteristic length can be taken as b for these numbers [127].

5.5 Water Jet Propulsion

Jet propulsion can be an appropriate strategy for aquatic escape, principally because it could produce near-continuous thrust as it transitioned from one medium to another.

Initial insight into key design considerations can be gained by considering an idealised water jet propelled robot, with empty mass m, expelling a finite volume of water V_w through a nozzle at constant pressure, P (Fig. 5.3). Neglecting viscosity and assuming that the upstream fluid is stagnant, this would result in a constant nozzle outlet velocity, u, of

$$u = \sqrt{\frac{2P}{\rho_w}} \tag{5.10}$$

by Bernoulli's equation, where ρ_w is the density of water. This system is an 'ideal rocket' and the Tsiolkovsky equation applies. This gives the change in the robot's velocity, Δv, after all water is expelled, in the absence of external forces:

$$\Delta v = u \ln\left(\frac{m + \rho_w V_w}{m}\right) \tag{5.11}$$

During take-off, the robot aims to achieve sufficient speed for transition to flight, so maximising Δv is a sensible design objective. Combining Eqs. 5.10 and 5.11 gives

$$\Delta v = \sqrt{\frac{2P}{\rho_w}} \ln\left(1 + \frac{\rho_w V_w}{m}\right) \tag{5.12}$$

from which it can be seen that the robot's performance can be improved either by raising the fluid pressure or by increasing the propellant mass, $\rho_w V_w$ relative to the empty mass. However, the empty mass (the superstructure and control systems of the device) will correlate with the system's volume. The most immediate correlation is structural; the device contains a volume under pressure, which must be sustained without failure. For a thin walled cylindrical pressure vessel, the vessel mass is directly proportional to the internal volume and pressure, such that

$$m = m_{comp} + C_1 P V_w \tag{5.13}$$

Fig. 5.3 Idealised water jet device nomenclature: a system of mass m expels a finite volume of water, V, at pressure P, resulting in a nozzle outlet speed of u and a forward velocity v

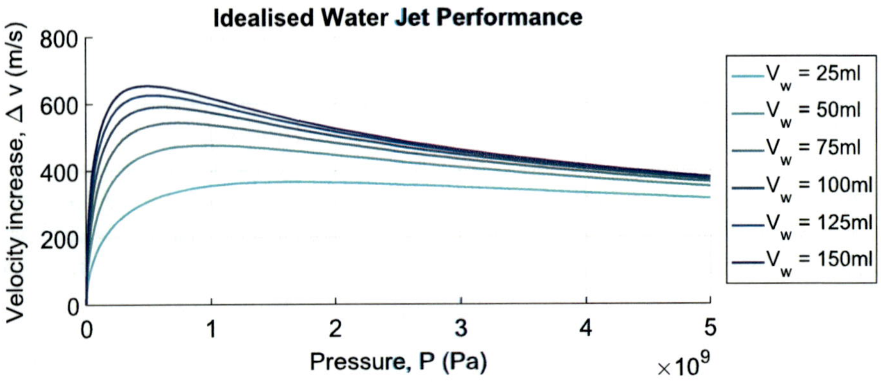

Fig. 5.4 Idealised water jet Δv against pressure, using the material properties of 7075 aerospace aluminium, plotted for several different vessel volumes. It can be seen that increasing pressure improves performance up to impractically large pressures before the function ultimately tends to zero as pressure goes to ∞

where m_{comp} is the mass of ancillary components and C_1 is a constant of proportionality determined by the density and yield stress of the pressure vessel material. This means that Eq. 5.12 becomes

$$\Delta v = \sqrt{\frac{2P}{\rho_w}} \ln\left(1 + \frac{\rho_w V_w}{m_{comp} + C_1 P V_w}\right) \tag{5.14}$$

and it can immediately be seen that Eq. 5.14 has a limit such that

$$\Delta v \rightarrow \sqrt{\frac{2P}{\rho_w}} \ln\left(1 + \frac{\rho_w}{C_1 P}\right) \qquad \text{where} \qquad V_w \rightarrow \infty \tag{5.15}$$

and there is a limit to the increase in performance that can be gained by increasing V_w. If pressure is instead increased, it can be shown that inn the limit of large P, Δv will tend to zero. However, there is an inflection point in the function prior to this, and an optimum pressure exists. This is examined here by prescribing a 50 g ancillary component mass and taking the material properties of 7075 aluminium (density 2.7 kg/m^3, yield stress 500 MPa), values reflective of prototypes presented in later chapters. If Δv is plotted against P, for several different vessel volumes (Fig. 5.4), it can be seen that this optimum occurs at pressures of 5×10^9 Pa, or 5,000 bar, far in excess of the pressures which can be practically achieved in a robot prototype. It can therefore be inferred that the most effective water jet device will be one which produces large pressures, whilst minimising the extra mass carried, and that for improving device performance it is preferable to increase pressure than to increase propellant volume.

Finally, Eq. 5.14 considers only the reachable speed of the device, and ignores the energetic cost of obtaining it. Since pressure is constant in this example, the work done in expelling the fluid is simply PV, and the velocity obtained per unit energy input is:

$$\frac{\Delta v}{PV_w} = \frac{1}{V_w}\sqrt{\frac{2}{P\rho_w}} \ln\left(1 + \frac{\rho_w V_w}{m_{comp} + C_1 P V_w}\right) \tag{5.16}$$

This means that a small, low pressure device is more energetically efficient, and that increasing the total initial energy of the system is more effective when achieved by an increase in pressure than an equivalent increase in volume.

Analysis of an idealised water jet system indicates that performance diminishes as size increases, and that it is thereby more suitable for use in small scale vehicles. The governing equations also indicate that the most effective way to improve performance is to increase jet pressure, which will guide the development of practical jet thruster systems.

5.6 Plunge Diving

A body entering the water at high speed experiences several decelerating forces; buoyancy, drag and resistance from the rapid change in fluid inertia or 'added mass'. An underwater vehicle is generally designed to be neutrally buoyant, such that it does not have to oppose gravity as it locomotes, so here discussion is limited to inertial forces and drag.

Considering first the effect of quadratic drag on a neutrally buoyant object descending through the water, a simple non-linear differential equation can be written:

$$m\frac{\mathrm{d}v}{\mathrm{d}t} = -\frac{1}{2}\rho_w C_D A v^2 \tag{5.17}$$

where m and v are again the object's mass and velocity, C_D is drag coefficient and A is a reference area. This can be readily solved given the boundary conditions $v(0) = v_0$, and $x(0) = 0$ (zero displacement at time $t = 0$), to give

$$v(t) = v_0\left(1 + \frac{\rho_w C_D A v_0 t}{2m}\right)^{-1} \tag{5.18}$$

$$x(t) = \frac{2m}{\rho_w C_D A} \ln\left(1 + \frac{\rho_w C_D A v_0 t}{2m}\right) \tag{5.19}$$

x is unbounded due to the way drag is defined here, which will stop being physical at low speeds, but the distance the object penetrates before it loses speed can nonetheless be examined by setting $v(t_{end}) = v_0/C_2$, where C_2 is an arbitrary constant:

$$x(t_{end}) = \frac{2m}{\rho_w C_D A} \ln(C_2) \tag{5.20}$$

$x(t_{end})$ then depends on $m/C_D A$, and since m is proportional to volume, the depth that can be achieved will increase as the object increases in size, suggesting that plunge diving is becomes an ineffective means of achieving depth as a robot is miniaturised.

However, if the body is not neutrally buoyant, and instead has an average density $\rho_B < \rho_w$, the equation of motion becomes

$$m\dot{v} = -\frac{1}{2}\rho_w C_D A v^2 - g V_B(\rho_w - \rho_B) \tag{5.21}$$

where V_B is the displaced volume and g is acceleration due to gravity. The distance the body will penetrate below the water is now bounded and is given by (see [128] for a full derivation):

$$x|_{v=0} = \frac{\rho_B V_B}{\rho_w C_D A} \ln\left(1 + \frac{\rho_w C_D A v_0^2}{2g V_B(\rho_w - \rho_B)}\right) \tag{5.22}$$

Now, if size is increased, V_B/A will increase linearly ($V_B/A \propto [L]^3/[L]^2 = [L]$), but the dive depth has a limit such that

$$x|_{v=0} \to \frac{v_0^2}{2g}\left(\frac{\rho_B}{\rho_w - \rho_B}\right) \quad \text{as} \quad \frac{V_B}{A} \to \infty \tag{5.23}$$

If, for example, a vehicle is intended to plunge dive to a certain depth, but passively return to the surface under buoyancy, this means it is better to increase depth by increasing impact speed (as $x \propto v_0^2$), or to decrease relative density, but increasing overall size has diminishing returns.

Dives into water are less effective at achieving depth as size is decreased, due to the unfavourable scaling of fluid drag, but a vehicle that is less dense than water can only achieve limited depth gains by increasing its size, and it is instead better to increase impact velocity, as long as the vehicle is slender enough to sustain inertial impact loads.

5.6.1 Water Impact

Increasing speed will in turn increase the loads that the object experiences when impacting the water, particularly the inertial loads. Fluid inertia forces arise from the mass of fluid which is also in motion around a travelling body, and the relative size of this mass depends principally on the shape of the object in motion. In air, it can be readily neglected due to the low medium density, but it becomes much more significant in water. For a body of mass m with added mass m_a, Newton's third law then becomes:

$$F = \frac{\mathrm{d}}{\mathrm{d}t}\Big((m + m_a)v\Big) = (m + m_a)\dot{v} + \dot{m}_a v \tag{5.24}$$

where F is the resultant external force. Importantly, the added mass forces are inviscid and conservative, such that for a fully submerged body the work done to accelerate the surrounding fluid will be recovered as inertia during deceleration, and so the presence of added mass should have minimal effect on final penetration depth during a dive. However, during impact at high speed the body entering the water accrues fluid inertia very rapidly, and the $\dot{m}_a v$ term becomes large, placing significant stress on the impacting body, which can place a structural limit on dive speed.

Von Karman developed an approximation for the added mass forces encountered during water impact [129], by approximating the instantaneous fluid inertia of the body as half that of a circular flat plate of the same area, such that

$$m_a = \frac{4}{3}\rho_w r^3 \tag{5.25}$$

where r is the instantaneous radius of contact with the water. This gives an expression for the inertial force an object experiences as it impacts the water

$$\dot{m}_a v = 4\rho_w r^2 \dot{r} v = 4\rho_w r^2 v^2 \left(\frac{\mathrm{d}r}{\mathrm{d}x}\right) \tag{5.26}$$

where x is again the depth below the water. This formulation has been shown to give accurate predictions of forces experienced by gannets impacting the water [130]. From the expression it can be seen that the water impact force has a very similar form to a quadratic drag, being proportional to frontal area (the r^2 term) and dynamic pressure (the $rho_w v^2$ term). The presence of the $\mathrm{d}r/\mathrm{d}x$ term indicates that a 'pointier' projectile will experience a lower impact force, an intuitive result.

To minimise drag, reductions in skin friction resulting from total wetted area must be traded against pressure drag resulting from projected frontal area. Under hydrodynamic drag, the fineness ratio (maximum length over maximum width) which minimises total drag for a given volume is 4.5 [131]. However, the presence of water impact forces means that there is an additional penalty attached to frontal area, and an appropriate shape for a plunge diving vehicle is more slender than that which minimises parasitic drag. This additional frontal area penalty will strongly affect the diving ability of a fixed wing vehicle, as a large wing area for low loading, with an efficient, high aspect ratio wing will necessarily have a large frontal area. This confirms the necessity of wing folding, not only to reduce drag but to protect the wing structure.

Part II
Practical Aerial Aquatic Locomotion

Chapter 6
Aquatic Escape: Fast Escape with a Jet Thruster

6.1 Compressed Gas Water Jet Thruster

In a previous chapter an idealised water jet thruster was analysed, and it was argued that the most effective system would use large pressures to drive a small volume of water. In this chapter a more detailed physical model of water jet propulsion will be introduced, and the key design features of a jet thruster prototype detailed. Consistent static thrust from the fabricated device is then demonstrated.

The presented thruster is intended to form a critical component of an AquaMAV able to plunge dive and relaunch itself, producing the thrust required reach to a velocity sufficient for flight from the water. To be used in sustained flight, the thruster must therefore be as lightweight as possible. During an impulsive take-off, the thruster will be actuated on the water surface and thrust production must not be sensitive to the variations in attitude due to the motion of the water surface. In order to be able discharge the device effectively at any angle, air and water must be kept separately, rather than free to mix in a single tank. Separate tanks also allows propellant water to be collected in situ at launch rather than carried, reducing flying weight. To achieve large pressures, the system was designed to charge from common liquid CO_2 capsules, which produce pressures of 57 bar at room temperature, in order to create a powerful, but practical and portable device. Liquid CO_2 capsules also produce gas at a constant pressure (the liquid vapour pressure), so a single capsule can recharge the device repeatedly without performance loss. The 16 g capsules used to charge the device are sufficient for 20 charges.

6.2 Water Jet Propulsion Theory

In this section, the underlying theory used in the design of the robot is introduced, including an sizing the water tank to maximise the useful thrust extracted from the gas tank. The subscripts 1, 2, 3 and 4 are used to denote variables relating to the main

R. Zufferey et al., *Between Sea and Sky: Aerial Aquatic Locomotion in Miniature Robots*, Biosystems & Biorobotics 29, https://doi.org/10.1007/978-3-030-89575-4_6

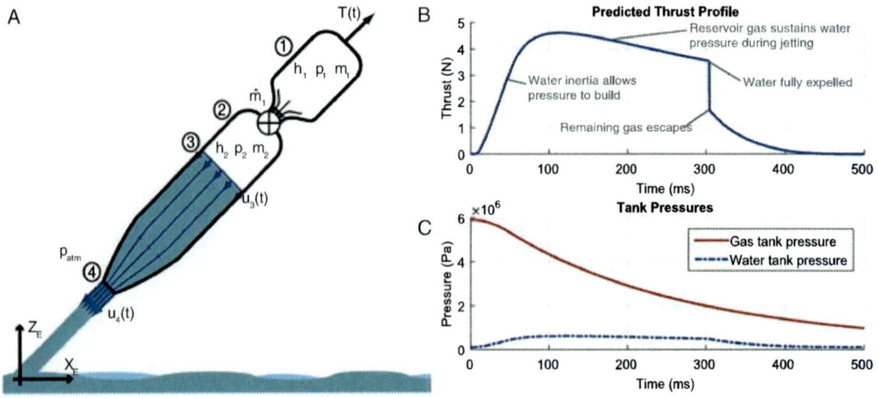

Fig. 6.1 (**A**) Jet propulsion principle: Gas released from a high pressure tank expels water, propelling the vehicle. Circled numbers correspond to the locations indicated by equation subscripts. (**B**) Simulated thrust for water rocket with separated chambers. (**C**) Gas pressures in water and gas tanks during jetting

gas tank, the gas within the water tank, the air-water interface, and the nozzle outlet respectively. Figure 6.1A illustrates nomenclature for the water jet flow.

The thrust produced by a jet of mass flow $\dot{m}_4$ and velocity u_4 is given by

$$T = \dot{m}_4 u_4 \tag{6.1}$$

If a gas is used as the propellant mass, its low density means that thrust production is negligible without very high exit velocities, and for efficient propulsion from a limited reservoir, a heavier propellant is preferable. For an AquaMAV, water is a propellant that can easily be collected before launch, with the expansion of compressed gas powering expulsion.

The incompressibility of water means the expelled jet will be at ambient pressure, and the gas expansion rate will equal the water outflow:

$$u_4 = \dot{V}_2 / A_4 \tag{6.2}$$

where u is the water velocity and V_2 the gas volume in the water tank. The water flow within the tank can be treated as quasi-1D by assuming uniform axial flow across the jet section [132]. By mass continuity, the local velocity is then a function of cross-sectional area:

$$A_3(t)u_3(t) = A_4 u_4(t) \tag{6.3}$$

where A_n the jet cross sectional area. The unsteady Bernoulli equation (Eq. 6.4) is recovered from Euler's equation by integrating from the air-water interface to the nozzle exit (Fig. 6.1). Total pressure along a streamline running from 3 to 4 is equal to the instantaneous gas pressure in the water tank:

$$\int_3^4 \frac{\partial u}{\partial t} ds + \frac{p_2}{\rho_w} + \frac{1}{2}(u_4^2 - u_3^2) = 0 \tag{6.4}$$

where p_2 is the pressure of gas in the water tank and ρ_w the density of water. The pressure acting on the water must be built up by the gas released from the CO_2 tank. To compute the flow rate out of the tank, the valve flow equations given in the European standard EN-60534 [133] are used:

$$\dot{m}_1 = K_v \Upsilon \sqrt{\kappa p_1 \rho_1} \tag{6.5}$$

$$\kappa' = (p_1 - p_2)/p_1 \tag{6.6}$$

$$\kappa = \begin{cases} \kappa' & \text{if } \kappa' < \kappa_{choke} \\ \kappa_{choke} & \text{if } \kappa' \geq \kappa_{choke} \end{cases} \tag{6.7}$$

With the gas tank initially charged to 57 bar, the outflow will be choked, and will remain so until the pressure ratio (Eq. 6.6) falls below κ_{choke} (Eq. 6.7). The mass flow depends on the valve flow coefficient, K_v, and the limiting pressure ratio, κ_{choke}, the point at which the flow through the valve becomes sonic. Υ is a compressibility correction factor [133]:

$$\Upsilon = 1 - \kappa/3\kappa_{choke} \tag{6.8}$$

Liquids and gases behave similarly at low pressures, so K_v was measured by fixing the valve in the open position and logging the discharged volume against time of a 0.5 m tall, 4 cm diameter column of water through the valve, and fitting K_v according to the EN-60534 [133] equations for incompressible fluids. κ_{choke} is a compressible flow property which cannot be measured from water flow, and so was inferred from manufacturer data for air flow at 7 bar based on the measured K_v value, and corrected for the different properties of CO_2. While measured at 7 bar, κ_{choke} is not expected to change at the larger initial pressures in the fabricated tank, although the higher upstream density will result in increased mass flow (Eq. 6.5).

To determine the variation of gas conditions in the two tanks during gas release, a first law energy balance is used. The gas exchange is treated as a quasi-equilibrium, adiabatic process, as jetting takes place over too short a timescale for significant heat transfer to occur. This gives an equation in which the stagnation enthalpy flux from the gas tank ($\dot{m}_1 h_{01}$) is equivalent to the increase in enthalpy and kinetic energy of gas in the water tank ($m_2(h_2 + u_3^2/2)$), minus the pdV work done against water pressure (Eq. 6.9). Gas conditions obey the ideal gas equation of state throughout.

$$\dot{m}_1 h_{01} = \frac{d}{dt}\left[m_2\left(h_2 + \frac{u_3^2}{2}\right)\right] - p_2 \dot{V}_2 \tag{6.9}$$

Where h is specific enthalpy (subscript $_0$ denotes a stagnation quantity). Combining Eqs. 6.2–6.9 leads to a system of four first and second order differential equations in $V_2(t)$, $\dot{V}_2(t)$, $h_1(t)$, $h_2(t)$ and $m_1(t)$. These equations remain valid until all water is expelled, after which point the release of the remaining gas produces a small amount of thrust. At this stage, total gas mass inside the thruster $(m_1 + m_2)$ is no longer conserved, Eq. 6.9 does not hold, and the mass flow out of the nozzle must also be included in the gas thermodynamic calculations (Eq. 6.10). To calculate mass flow out of the nozzle, the outlet Mach number, M, is calculated based on the ratio of outlet pressure to upstream stagnation pressure, p_{02} (Eq. 6.11).

$$\dot{m}_1 h_{01} = \frac{d}{dt}\left[m_2\left(h_2 + \frac{u_3^2}{2}\right)\right] - \dot{m}_4 h_{02} \tag{6.10}$$

$$\frac{p_4}{p_{02}} = \left(1 + \frac{\gamma - 1}{2} M^2\right)^{-\frac{\gamma}{\gamma-1}} \tag{6.11}$$

Where γ is the gas adiabatic index. The conical water nozzle has no diverging section so $M \leq 1$. When subsonic, the nozzle outlet will be at atmospheric pressure, but if the pressure ratio is greater than a critical value (Eq. 6.12, different to κ_{choke}, which is specific to the valve), the flow is choked, $M = 1$, and the nozzle outlet pressure will be greater than atmospheric. In both cases, the mass flow out can be computed using a standard Mach number relation (Eq. 6.13).

$$\frac{p_{atm}}{p_2} = \left(\frac{2}{\gamma + 1}\right)^{\frac{\gamma}{\gamma-1}} \tag{6.12}$$

$$\frac{\dot{m}_4\sqrt{c_p T_{02}}}{A_4 p_{02}} = \frac{\gamma M}{\sqrt{\gamma - 1}}\left(1 + \frac{\gamma - 1}{2} M^2\right)^{-\frac{1}{2}\frac{\gamma+1}{\gamma-1}} \tag{6.13}$$

Where c_p the gas heat capacity and p_{02} is the stagnation pressure of gas in the water tank). Thrust is given by Eq. 6.14 with an additional term added to account for the nozzle outflow being above atmospheric pressure.

$$T = \dot{m}_4 u_4 - A_4(p_4 - p_{atm}) \tag{6.14}$$

This system of equations is solved in Matlab. The rapidly varying fluid inertia term (Eq. 6.4) can cause numerical instability, so a variable order implicit solver is used. A conditional statement links the regimes discontinuously; integration of the water jetting equations is halted once all water is expelled, and final values provide initial conditions to the gas-only equations.

The simulated results for the prototyped thruster are shown in Fig. 6.1. Initially, the inertia of the fluid restricts the water expulsion, and allows pressure to be built up in the water tank. A small amount of additional gas thrust can be seen after the water is fully expelled at 0.3 s, reducing rapidly. Due to the high initial pressures, the gas flows through the valve and nozzle are choked throughout jetting.

6.2.1 Design Domain

For a given reservoir pressure and valve flow coefficient, the amount of work extracted from the gas can be maximised by varying the capacity of the water tank and the nozzle outlet diameter. Enlarging water tank capacity increases launch mass, so there will be an optimum tank volume. To obtain this optimum, the specific total impulse (I_{sp}, Eq. 6.15) is used as an objective to maximise the momentum imparted to the thruster.

$$I_{sp} = \int T dt / m_{total} \tag{6.15}$$

During jetting, pressure in the water tank is maintained by gas from the reservoir with a limited flow rate, so a smaller nozzle will allow a higher pressure to be maintained as the gas expands, increasing performance. However, this decreases the jet thrust (Eq. 6.1) and a very small nozzle will be insufficient to propel the vehicle. It was decided to target <5 N of peak thrust (a thrust to weight ratio greater than 5 for a 100 g robot), giving a minimum nozzle exit diameter of 2 mm, rounded for manufacturing. The design domain was computed by numerical integration of the equations in Sect. 6.2 (Fig. 6.2), with the specific impulse calculated based on the mass of the thruster alone, excluding the electronics and airframe. This gave an optimum tank length of 0.45 m, based on the 11 mm diameter tube used to form the water tank, which was fabricated.

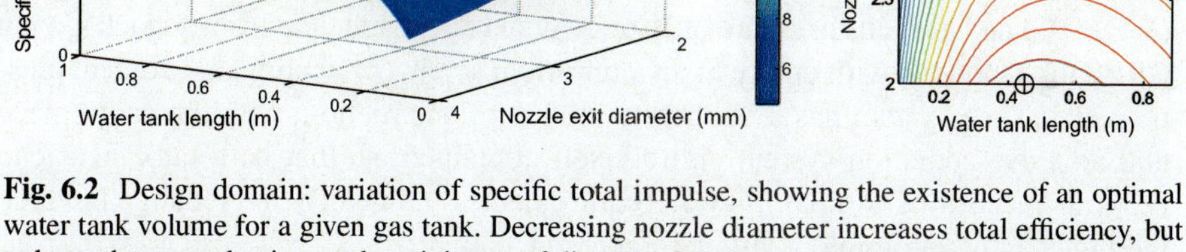

Fig. 6.2 Design domain: variation of specific total impulse, showing the existence of an optimal water tank volume for a given gas tank. Decreasing nozzle diameter increases total efficiency, but reduces thrust production, and a minimum of diameter 2 mm was set. The prototyped geometry is marked with a ⊕

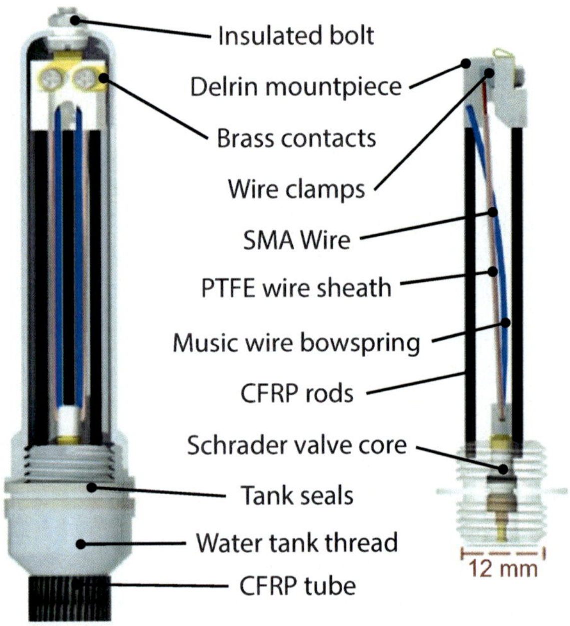

Fig. 6.3 Close up of the gas tank and the self contained, removeable SMA gas release system. The SMA wire is coloured red, while the buckling spring steel wires are shown in blue

6.3 Prototype

The fabricated thruster consists of an air and water tank, with sealed screw connections to a centrepiece containing a poppet valve core (Fig. 6.3). To reuseably contain and release the high pressure gas, a release valve controlled by NiTi Shape Memory Alloy (SMA) wires has been developed. This valve actuator is inside the CO_2 tank, so canisters can be easily attached for charging (Fig. 6.3).

The gas pressure vessel is constructed from 7075 aluminium according to European standards [133], with an additional safety factor of 2 applied to the wall thickness due to an abundance of caution. The water tank is pressurised to less than 10bar (Fig. 6.1C) and sustains pressure only briefly, so is instead made from a stock woven CRFP tube, bonded with epoxy to an aluminium screw connection and conical plastic nozzle. The system has a deliberate modular construction, with the centrepiece and its valve actuation system entirely self contained, so that both tank sizes can be varied according to final mission requirements, and readily interchanged, which facilitates integration into a larger vehicle in Chap. 7.

6.3.1 *Valve Actuation*

The valve is opened by moving the valve stem 1.6 mm vertically. When under 57 bar of internal pressure, the force required to open the valve is 24 N, (19 N of pressure

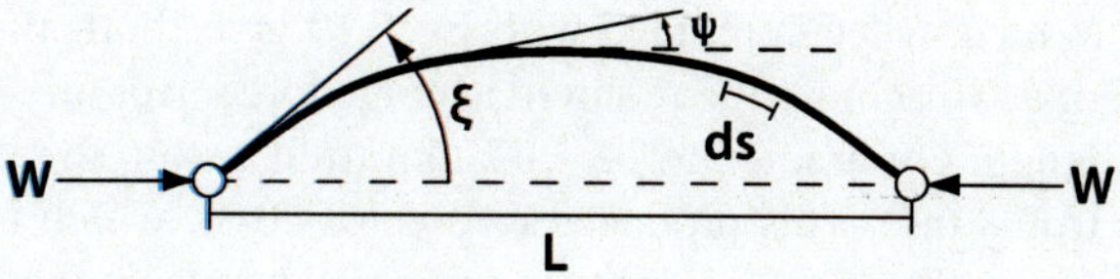

Fig. 6.4 Elastica problem definition for the buckled bars used antagonistically with the SMA wire

force, 5 N from the valve internal spring). These high force, short stroke actuation requirements are well suited to the use of NiTi SMA wire, and an actuation system was designed using 0.51 mm wire. These wires produce repeatable contraction forces of up to 35.6 N upon heating to 90°C. To provide sufficient stroke and force, a single length of 104 mm flexinol wire is threaded through the valve stem, creating two 50 mm wires connected in series electrically, but acting in parallel mechanically (Fig. 6.3), giving a combined force of up to 71.2 N. The wires pull against a frame formed by carbon fibre rods mounted into the centrepiece.

After contraction, the wires must be stretched to their original length to reseal the tank. This requires a minimum stress of 69 MPa, corresponding to a force of 27.2 N. To achieve this force with minimal mass and volume, two buckling spring steel (music wire) bars with pinned ends were used. The governing equation for the large deflection buckling behaviour of a pinned beam compressed from undeformed length L_0 to length L is, in terms of local rotation ψ and a co-ordinates s along the beam length (Fig. 6.4):

$$EI\frac{\mathrm{d}^2\psi}{\mathrm{d}s} = -W\sin\psi \tag{6.16}$$

Where E is the flexural modulus of the material, I is the second moment of area, and W is the resultant force at the hinges (Fig. 6.4). With the boundary condition $\psi''|_{s=0} = \psi''|_{s=L_0} = 0$. An analytic solution to this equation exists, and can be expressed using elliptical integrals [134]:

$$\frac{W}{W_{eu}} = \frac{4K^2}{\pi^2} = \frac{4}{\pi^2}\left(\int_0^{\frac{\pi}{2}} \frac{\mathrm{d}\phi}{\sqrt{1-\sin^2\frac{\xi}{2}\sin^2\phi}}\right)^2 \tag{6.17}$$

where ξ is the deflected angle of the beam at the pin ($\xi = \psi|_{s=0}$, Fig. 6.4). K is the complete elliptic integral of the first kind, and W_{eu} denotes the Euler buckling load:

$$W_{eu} = \frac{\pi^2 EI}{L_0^2} \tag{6.18}$$

The length between the pins of the deformed beam L, is given by:

$$L = 2K\sqrt{\frac{EI}{W}} \tag{6.19}$$

Which, when combined with Eq. 6.19 gives both W and L parametrically in terms of ξ. To achieve the required average force over the valve stroke, with the spring bar length constrained to be <47 mm to fit within the tank, the wire diameter was set to 0.8 mm. This produced a force of 40 N, which is near-constant over the stroke range. The steel is slightly prebuckled to ensure that the wires deflect in the correct direction when the valve opens.

To provide an electrical conduit into the pressurised container, the vessel wall is used as a negative earth and an insulated bolt was fastened through the tank end (Fig. 6.3). Brass contacts are attached to each end of the SMA wire, contacting the vessel wall and bolt when the valve assembly is inserted, forming a complete circuit without compromising the seal. The SMA wires are actuated by a 100 Hz, 9.1 A current pulse train with 50% duty cycle, lasting for 1 s. This requires 1.3 mAh of charge from a 7.4 V battery. The current pulse is controlled and timed using an Arduino microcontroller. Excluding control electronics, the mass of the thruster is 41.9 g.

In order to leap clear of the water, the jet tank must ingest water from its surroundings. While the water tank will fill gradually with the tail pointing upward in the water, it will not fill when nose up. For testing purposes the water tank was initially filled manually before actuation, but the prototype in Chap. 7 incorporates a check valve at the opposite end to the water tank nozzle, allowing the tank to passively fill with water before being pressurised by gas release, but containing the pressure when the jet is actuated.

6.3.2 *Static Performance*

The device's static thrust was measured by mounting it vertically to a load cell, with force data recorded at 2500 Hz (Fig. 6.5). The sensor was zeroed with the jet water tank full, so the weight reduction as water is expelled was also measured by the force sensor. The model output has been modified to show this effect in the predicted force profile shown in Fig. 6.5. Thrust profiles were recorded for six actuations, with the final actuation conducted with the jet immersed in water. The measured

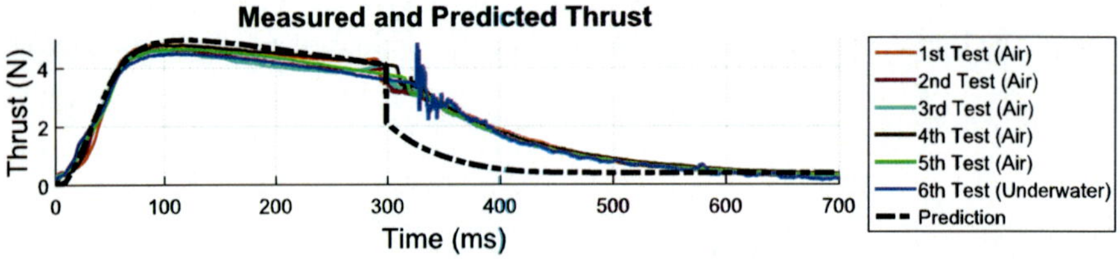

Fig. 6.5 Thrust test data. Slight inconsistencies in thrust are due to variations of the vapour pressure of the liquid CO_2 in the charging capsule. Test 6 was conducted with the thruster immersed in water. The average of the profiles shown is used in trajectory simulations in Sect. 6.4

thrust profiles show the expected features of a rapid rise in pressure as the water accelerates, followed by water discharge at sustained pressure and a tapering off of thrust as remaining gas is expelled.

The measured thrust profiles show good consistency between repetitions. The slight discrepancies observed are likely due to a change in the temperature of the CO_2 tank used for charging. The vapour pressure of CO_2 is 57 bar at 20 °C, but this varies by approximately 1.25 bar/°C, and both the cooling of the capsule due to liquid CO_2 evaporation during charging and changes in ambient conditions will slightly vary initial pressure between tests.

Towards the end of the water expulsion, the velocity of the air-water interface increases rapidly as it passes through the nozzle contraction. At this point, fluid wall shear and 'plughole' vortical instabilities allow air to mix with water and escape before all water has been expelled, resulting in a spray rather than a jet. The effect of this spraying can be observed in the tapering thrust profile in Fig. 6.5, compared to the abrupt drop and brief gas-only discharge predicted by theory (Fig. 6.5). The observed thrust begins to taper earlier than the predicted discharge end, and increased noise can be observed from the spraying. However, this also impedes the gas flow out, and allows more gas to be built up from the reservoir tank than is predicted by theory, and thrust production continues for longer. This means that the total impulse is not greatly changed from prediction. No significant variation in thrust when actuating underwater was expected, which was confirmed by the sixth static thrust test, conducted underwater. The only appreciable difference was the increased noise in the thrust profile after air begins to mix with the water outflow and the air jet forms into bubbles.

6.4 Underwater Takeoff Using a Jet Thruster

In this chapter an AquaMAV capable of gliding leaps from beneath the water is presented. The robot leaps clear of the water using the water jet detailed in Sect. 6.1, driven by the controlled release of a small 5 ml tank of 57 bar CO_2 gas. The robot uses a shape memory alloy actuated valve to control the release of stored CO_2, and is fitted with a pair of deployable wings, which allow it to maintain stability and minimise drag when leaving the water. These wings are then deployed in the air for gliding. A planar trajectory model of transition across the water surface is used to examine the mechanics of aquatic take-off, and demonstrate that an impulsive jet is a robust means of transition to flight. Finally, flight from beneath the water is then demonstrated.

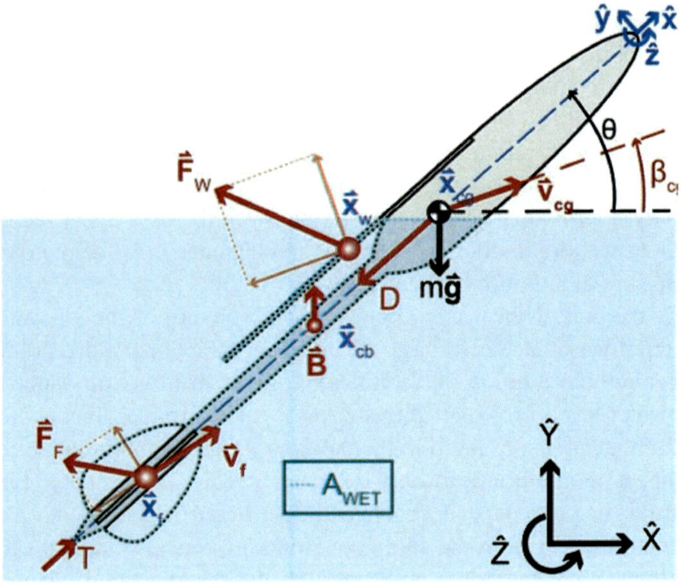

Fig. 6.6 Nomenclature for the equations of motion

6.5 Planar Trajectory Model

The jet is given wireless control, and is integrated with collapsible wings and tail fins for flight. To investigate the robustness of the transition to flight from water, a planar trajectory model is implemented, with a simple estimation of the hydrodynamic forces during water exit (Sect. 6.5.3). Here, the subscripts $_w$, $_f$, $_b$, $_{cg}$ and $_{cb}$ are used to refer to the robot wing, fins, body and centres of gravity and buoyancy respectively. Subscript $_s$ refers to skin friction forces. The trajectory is defined by velocity and acceleration vectors ($\vec{a}$ and $\vec{v}$ respectively) in earth fixed inertial axes with unit vectors $\hat{X}$, $\hat{Y}$, $\hat{Z}$. Position vectors, $\vec{x}$, are also defined within a robot-fixed reference frame rotated by an angle θ about $\hat{Z}$ with respect to the in inertial frame, with its origin at the robot nose and with unit vectors $\hat{x}$, $\hat{y}$, $\hat{z}$ (Fig. 6.6).

6.5.1 Aerodynamics

The vector forces produced by the wing, $\vec{F}_w$, and the tail fins, $\vec{F}_f$ (Eqs. 6.20 and 6.21), have a lift component perpendicular to the local velocity, $\vec{v}$, and a drag component opposing it, defined by lift and drag coefficients C_{lw} and C_{dw} respectively for the main wing, and C_{lf} and C_{df} for the fins:

$$\vec{\boldsymbol{F}}_{\boldsymbol{w}} = \frac{1}{2}\begin{pmatrix} C_{lw} \\ C_{dw} \end{pmatrix} \rho_a A_w |\vec{\boldsymbol{v}}_{cg}|^2 \tag{6.20}$$

$$\vec{\boldsymbol{F}}_{\boldsymbol{f}} = \frac{1}{2}\begin{pmatrix} C_{lf} \\ C_{df} \end{pmatrix} \rho_a A_f |\vec{\boldsymbol{v}}_f|^2 \tag{6.21}$$

Where A is the component's lifting area, and ρ_a the density of air. The local velocity of the wing is taken as the velocity of the centre of gravity, $\vec{\boldsymbol{v}}_{\boldsymbol{cg}}$, but the tail velocity, $\vec{\boldsymbol{v}}_f$ includes the effect of the vehicle's pitch rate, $\dot{\theta}$:

$$\vec{\boldsymbol{v}}_f = \vec{\boldsymbol{v}}_{cg} + \boldsymbol{R}(\pi/2 - \theta)(\dot{\theta}\hat{\boldsymbol{Z}} \times \vec{\boldsymbol{x}}_f) \tag{6.22}$$

Where $\boldsymbol{R}(\pi/2 - \theta)$ is a matrix representing rotation by $\pi/2 - \theta$ about $\hat{\boldsymbol{Z}}$. The main wing is initially retracted into a low aspect ratio ($_w$) delta configuration (Fig. 6.7) which will produce significant vortex lift, and so the Polhamus suction analogy is used to compute C_{lw} and C_{dw} (Eqs. 6.23 and 6.24, with constants k_p and k_v taken from [135]). In the model, the deployment of the wings is treated as instantaneous, represented by a step change in the wing's lift behaviour, at time $t = t_{deploy}$:

$$\begin{aligned} C_{lw} =& k_p sin(\boldsymbol{\alpha}_w) cos^2(\boldsymbol{\alpha}_w) \\ &+ k_v sin^2(\boldsymbol{\alpha}_w) cos(\boldsymbol{\alpha}_w) \end{aligned} \tag{6.23}$$

$$C_{dw} = k_p sin^2(\boldsymbol{\alpha}_w) cos(\boldsymbol{\alpha}_w) + k_v sin^3(\boldsymbol{\alpha}_w) \tag{6.24}$$

$$\text{while: } t < t_{deploy} \tag{6.25}$$

Once open, the wing has a high aspect ratio, and C_{lw} and C_{dw} become the coefficients of an elliptic flat plate (Eqs. 6.26 and 6.27):

$$C_{lw} = 2\pi(\boldsymbol{\alpha}_w)/(1 + 2_w^{-1}) \tag{6.26}$$

$$C_{dw} = C_{lw}^2/(\boldsymbol{\pi}_w) \tag{6.27}$$

$$\text{while: } t \geq t_{deploy}$$

The fin coefficient equations, C_{lf} and C_{df} have the same form as Eqs. 6.26 and 6.27. The angles of attack of the main wing and fins, $\boldsymbol{\alpha}_w$ and $\boldsymbol{\alpha}_f$, are calculated from the angle the component's velocity in the inertial frame makes with the horizontal, denoted by $\boldsymbol{\beta}$. $\boldsymbol{\beta}_f$ includes pitch damping effects (Eq. 6.22), while pitching of the main wing is neglected:

$$\boldsymbol{\alpha}_w = \boldsymbol{\theta} - \boldsymbol{\beta}_{cg} + \boldsymbol{\alpha}_0 \tag{6.28}$$

$$\boldsymbol{\alpha}_f = \boldsymbol{\theta} - \boldsymbol{\beta}_f \tag{6.29}$$

α_0 is the preset wing incidence relative to the robot's longitudinal axis, fixed at 7°. When deployed, the main wing has an area of 475 cm^2, a span of 450 mm and aspect

Table 6.1 Key robot parameters

Parameter	Value		Unit
	Retracted	Deployed	
Wing area, A_w	126	475	cm^2
Wing aspect ratio, $_w$	0.79	4.26	–
Wing span	100	450	mm
Wing chord	210	132	mm
Wing preset angle of attack, α_0	7		^{o}
Tail area, A_f	50		cm^2
Tail aspect ratio, $_f$	1.4		–
Body wetted surface area, A_b	278		cm^2
Polhamus constant, k_P [135]	0.7		–
Polhamus constant, k_V [135]	3		–
Body width, BW	36		mm
Body length, BL	552		mm

ratio of 4.26 (Table 6.1), with an inverse Zimmerman planform. This means that with no incidence on the body and tail fins the AquaMAV must travel at 8.5 m/s to produce sufficient lift to balance its 100 g empty weight.

6.5.2 Longitudinal Stability

The location of the jet centre of mass, $\vec{x}_{cg}$, and the moment of inertia about that centre, I_{zz}, will change as water is expelled from the tank. In the simulation, the water remaining in the tank is calculated from experimental thrust data using Eq. 6.1. The motion of the centre of mass is of particular importance to the transition to flight, as it has a significant effect on the longitudinal aerodynamic stability. The retraction of the wings ensures stability during launch by moving the aerodynamic centre of the main wing, $\vec{x}_w$, rearward (Fig. 6.7). Once all water is expelled, $\vec{x}_{cg}$ is slightly ahead of the deployed wing quarter chord, and the wings can deploy into a larger, higher aspect ratio configuration for flight.

6.5.3 Water Resistance

The above equations for C_l and C_d neglect the effect of friction, so an additional drag force, D is included (Eq. 6.30) aligned with the robot's longitudinal axis. The jet body is slender and streamlined (Fig. 6.7), and the fins and wings are made from thin flat plates, so the drag at small angles of attack will be dominated by skin friction forces. In order to estimate the viscous force on the wings and fins, a turbulent flow

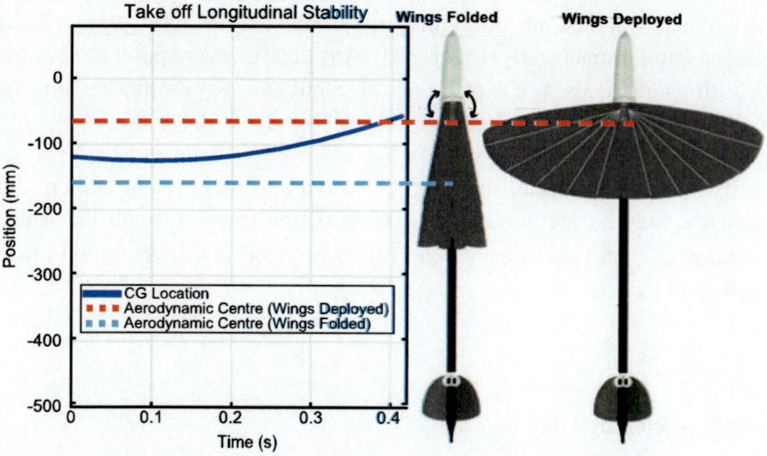

Fig. 6.7 Movement of the AquaMAV centre of gravity as water is expelled during jetting, with the aerodynamic centre of the wings in both configuration shown. Folding the wings moves the aerodynamic centre backward, ensuring stability until all water has been expelled

flat plate skin friction coefficient is used, using a Reynolds number (Re) based on retracted wing cord to estimate friction on the lifting surfaces (Eq. 6.31). To compute the body drag this coefficient is modified based on the ratio of the body's maximum width and length (BW/BL, Eq. 6.32) [136]. D is assumed to act through the robot's centre of gravity, in the negative $\hat{x}$ direction. The drag relations therefore become:

$$D = \frac{\rho_a}{2}(C_s(2A_w + 4A_f) + C_{sb}A_b)|\vec{v}_{cg}.\hat{x}|\vec{v}_{cg}.\hat{x} \tag{6.30}$$

$$C_s = 0.0307Re^{-1/7} \tag{6.31}$$

$$C_{sb} = C_s\left(1 + \frac{3}{2}(BW/BL)^{\frac{3}{2}} + 7(BW/BL)^3\right) \tag{6.32}$$

The robot is actuated while floating on the surface of the water, and has slight positive buoyancy, such that its nose just breaks the surface. As it accelerates out of the water, drag acting on the vehicle will limit its velocity. At each timestep, the model calculates the fraction of each surface that is immersed in the water, based on the robot position and orientation. Neglecting Reynolds number changes, the only fluid specific variable in Eqs. 6.20–6.32 is the fluid density. A factor, Q can therefore be introduced:

$$Q = \left(\frac{\rho_w}{\rho_a}\frac{A_{wet}}{A_{total}} + \left(1 - \frac{A_{wet}}{A_{total}}\right)\right) \tag{6.33}$$

Where A_{wet} and A_{total} are the submerged and total areas of the component in question. This approximation neglects any motion of the water surface, which is treated as remaining flat. It also assumes that the flow around a partially immersed surface is similar to the flow at the same point on a fully immersed surface (a similar assumption is made for water impact analysis in [137]). Any change in the point of action of these forces during aquatic escape is neglected. Despite these assumptions, this approach is capable of producing good predictions of speeds as the robot leaves the water (Sect. 6.8).

Finally, because the robot floats on the water surface prior to launch, the effect of buoyancy must be included. The volume of the robot beneath the water, V_{wet}, is calculated at each timestep, which gives a buoyant force, $\vec{\boldsymbol{B}}$, which always acts vertically, and its point of action, $\vec{\boldsymbol{x}}_{cb}$:

$$\vec{\boldsymbol{B}} = \rho_w g V_{wet} \hat{\boldsymbol{Y}} \tag{6.34}$$

where g is acceleration due to gravity).

6.5.4 Equations of Motion

To compute trajectories, the forces acting on the robot are resolved into inertial axes, and moments are taken about the jet centre of gravity, $\boldsymbol{x}_{cg}$:

$$\begin{aligned} m\vec{\boldsymbol{a}} &= \vec{\boldsymbol{B}} - m\vec{\boldsymbol{g}} + \boldsymbol{R}(\theta-\alpha_w) Q_w \vec{\boldsymbol{F}}_w \\ &+ \boldsymbol{R}(\theta)(T - DQ_s)\hat{\boldsymbol{x}} + \boldsymbol{R}(\theta-\alpha_f) Q_f \vec{\boldsymbol{F}}_f \end{aligned} \tag{6.35}$$

$$\begin{aligned} I_{yy}\ddot{\theta}\hat{z} &= (\vec{\boldsymbol{x}}_w - \vec{\boldsymbol{x}}_{cg}) \times \boldsymbol{R}(\alpha_w) Q_w \vec{\boldsymbol{F}}_w \\ &+ (\vec{\boldsymbol{x}}_{cb} - \vec{\boldsymbol{x}}_{cg}) \times \boldsymbol{R}(\theta)\vec{\boldsymbol{B}} \\ &+ (\vec{\boldsymbol{x}}_f - \vec{\boldsymbol{x}}_{cg}) \times \boldsymbol{R}(\alpha_f) Q_f \vec{\boldsymbol{F}}_f - \dot{I}_{yy}\dot{\theta}\hat{z} \end{aligned} \tag{6.36}$$

The thrust, T, used for the integration is the mean of 6 experimental thrust profiles (Sect. 6.3.2). The mass of water remaining in the tank is calculated from the experimental data using Eq. 6.1. The resulting equations of motion are integrated numerically to compute trajectories, using a 4th order Runge–Kutta solver in Matlab.

6.6 Take-Off Robustness

The simple drag model was found to give a good prediction of the acceleration profile of the AquaMAV during aquatic take-off (Sect. 6.8). The model was therefore used to evaluate the robustness of take-off to external perturbations which can occur in an outdoor environment. Wave motion over the floating vehicle can perturb the

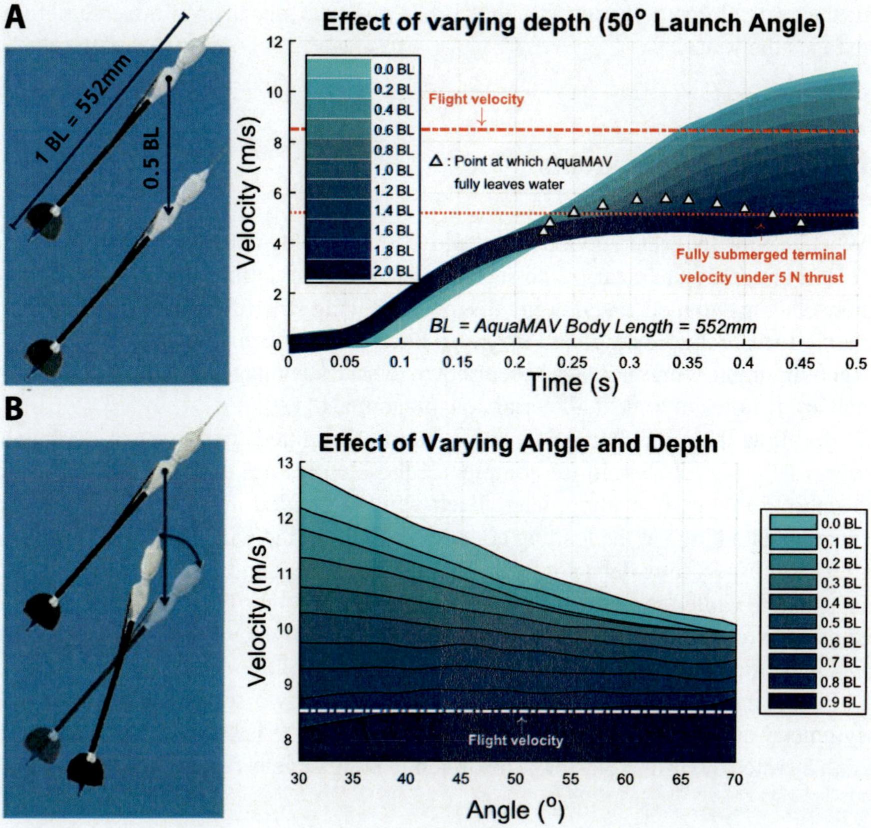

Fig. 6.8 The effect of perturbations on aquatic take-off. **A** The effect of increasing the depth of the robot beneath the water (50° start angle). Depth is expressed in body lengths (BL) below the natural floating position. The emergence speed is limited by drag, but the robot can achieve flight speed when submerged by up to 1 BL. **B** The effect of both depth and launch angle on the robot velocity after jetting. At up to 0.8 BL the robot can reach flight speed in spite of perturbations to initial launch angle

vehicle angle when it is actuated, and instantaneously submerge the robot, creating an additional distance to traverse underwater.

Increasing depth has a strong effect on the final velocity, as drag greatly limits underwater speed (Fig. 6.8A). Equating the drag (Eq. 6.30) when the robot is fully submerged with wings folded with the 5 N peak thrust gives a terminal velocity of 5.2 m/s (neglecting lift and buoyancy), which the robot velocity after jetting tends toward as depth increases. Simulating launches at several different angles and depths, the simulation indicates that the jet will be able to achieve its minimum flight velocity (8.5 m/s) regardless of angle as long as it is not submerged more than $0.8BL$ beneath the surface (Fig. 6.8B). Launching the robot nearer to vertical results in lower speed (but higher altitude). More interestingly, a steeper launch makes the robot less sensitive to the vertical depth change, because it has a shorter distance to traverse before it escapes the water.

6.7 Flight Components

When leaving the water, both flying squid [78] and flying fish [138] keep their wings folded until they are clear of the surface. There are large differences in fluid forces between the two media, and doing this protects wing structures from large hydrodynamic loads, reduces drag, and may also have stability considerations (Sect. 6.5.2). Reconfigurable wings have also been shown to have advantages in jumpgliding [101], and are features of almost all aerial aquatic animals [139].

To allow the aircraft to glide, the AquaMAV is fitted with a set of deployable wings (Fig. 6.9). To minimise complexity these wings are uncambered, fabricated from 0.25 mm carbon fibre plates. Each wing is divided into six segments which pivot about a hinge at the leading edge, which allows the wings to sweep backward into a low aspect ratio delta configuration. Of the six segments, only the leading edge is actuated, while the root segment is fixed to the fuselage. The other four segments are free to rotate, but are linked to the driven leading edge by kevlar twine, such that when the leading edge is rotated the other segments follow.

The wing hinges are actuated using 4.5 g servos. Two servos are used for mass symmetry and simplicity. The servos are attached to the jet water tank and covered with a sealed 3D printed faring. The driven leading edge is connected to the internal

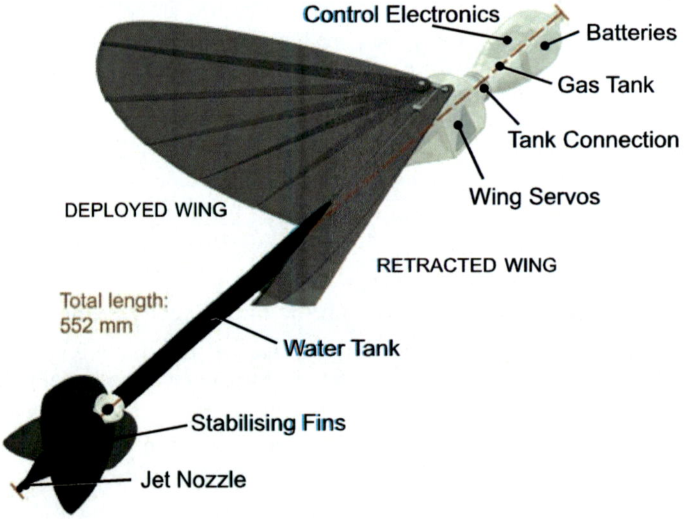

Fig. 6.9 CAD rendering of the fabricated AquaMAV prototype, showing component placement within the AquaMAV and the two wing configurations

Table 6.2 Robot weight breakdown

Component	Material	Weight [g]
Gas tank	Aluminium	12.3
Water tank	Aluminium/CFRP	21.6
SMA valve	Various	8.0
Electronics	Various	7.2
Batteries	7.4 V, 200 mAh Lipo	8.3
Wing servos	Various	8.9
Main wings	CFRP	19.5
Fuselage	PVC/ABS plastic	10.6
Tail fins	CFRP	4.4
Total mass		**100.8**

servos through a sealed bearing. The control electronics are contained in a separate fuselage section attached to the water tank which allows the gas tank to be removed easily for charging. Finally, the vehicle is fitted with four unactuated tail fins for stability, located at the end of the carbon tube. The total mass of the robot is 100.8 g (Table 6.2).

6.8 Aquatic Takeoff Performance

Once the static thrust was recorded, the AquaMAV was launched from a water tank into flight. The jet has slight positive buoyancy, and so was allowed to float freely on the surface. To validate the trajectory model, the launch angle was varied by resting the robot tail on submerged platforms of varying height. Trajectories were tracked using 800 fps video. The jet was directed to impact a net 4.5 m away from the launch point, which did not allow for a significant gliding phase to be recorded. To allow longer trajectories, the AquaMAV was also launched from a nearby lake and filmed, although tracking was not possible.

The indoor tracked tests showed that the trajectory model was able to accurately predict the AquaMAV's take-off (Fig. 6.10A), and demonstrated that the jet could readily achieve speeds of over 11 m/s, more than enough for transition to flight. The model predicts that the delta wing will generate significant hydrodynamic lift when in the water, which was captured by the model (Fig. 6.10B), as was the overall trajectory of the robot in several tests.

During outdoor tests (Fig. 6.11), the wings were deployed after a preset delay (t_{deploy}, Fig. 6.12A), determined using the trajectory model, based on maximising the total glide distance. Using this simple open loop method only limited gliding was achieved, and slight variations in trajectory resulted in the robot having insufficient time to passively correct its pitch before hitting the ground (Fig. 6.11B). The use of

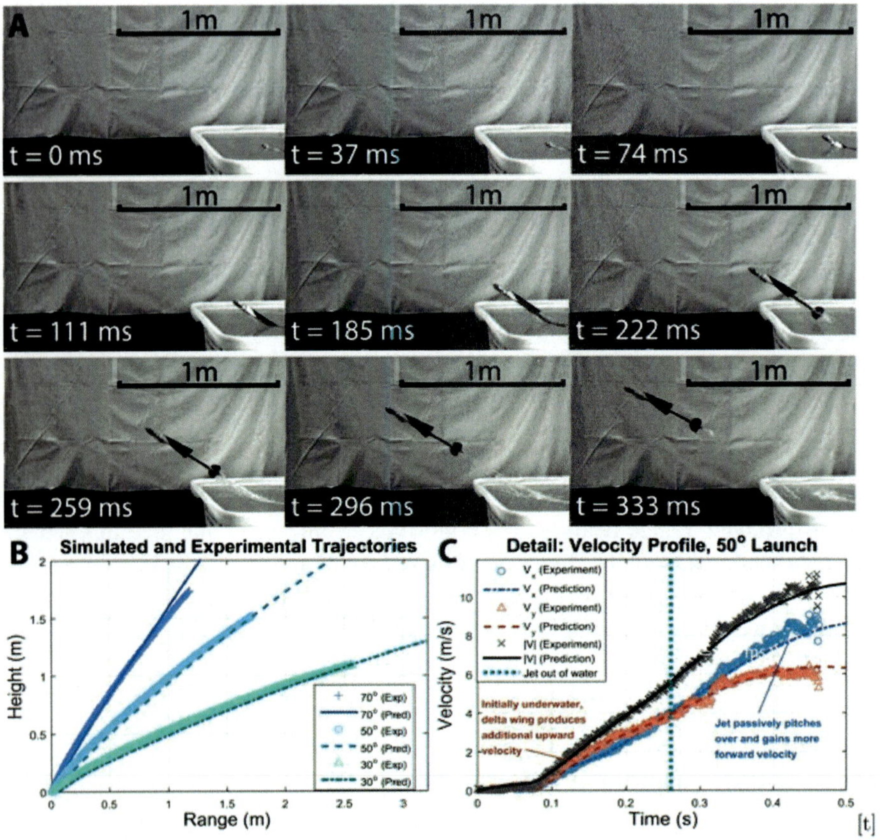

Fig. 6.10 Aquatic take-off tests. (**A**) Cropped stills from 800 fps underwater launch video (30° launch angle), 37 ms apart. (**B**) Three tracked trajectories, at 30°, 50° and 70°. (**C**) Velocity profile from a 50° launch, showing vertical and horizontal components. The nose of the jet is tracked so that the entire take-off can be analysed. The model is modified to output the position and velocity of this point, rather than the centre of gravity

Fig. 6.11 An AquaMAV can return water samples and data from isolated areas of water, using a powerful burst of water jet thrust to accelerate free of the water and transition to flight. **A** AquaMAV launches itself out of water. **B** Timelapse of a launch trajectory. Wings are deployed in final snapshot

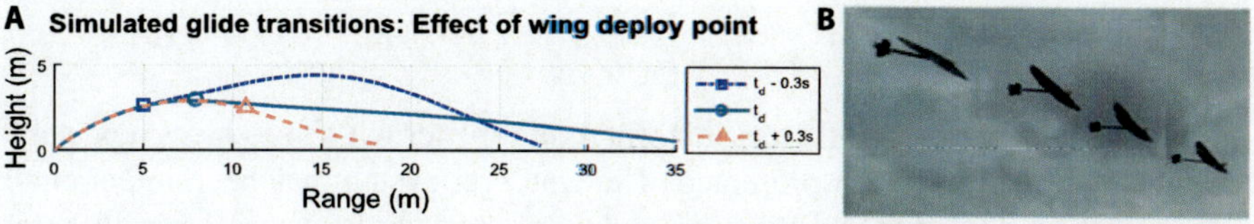

Fig. 6.12 **A** Trajectory simulations for a 50° launch with glide transition, with the wing deploy point indicated. In the absence of elevator control, deploying the wings early or late significantly effects the trajectory. **B** Uncommanded shifting of the left wing after deployment during a hand launched flight, leading to a crash

separately actuated wings often led to asymmetry between the two wings, rolling the robot (Fig. 6.12B). In addition, the direct link between servo and wing makes the wings prone to uncommanded movement; adding bistability to the mechanism would improve wing rigidity and performance.

The fabricated system has demonstrates the viability of the design and the reliability of the underlying theory. The robot is able to achieve significant flight speed from water in spite of hydrodynamic resistance, and a simple drag model has been shown to provide good predictions of launch trajectories. This model indicates that the robot is able to take-off in the presence of perturbations from surface waves to both its launch angle and depth.

Further work is needed to implement a reliable transition to gliding flight. Principally, effective take off will require some control, as even though flight speed is attainable despite launch perturbations, deploying wings and transitioning to stable, level flight will additional attitude correction. Some autonomy is also a necessary part of launching from the water because of communication limitations. The radio frequencies commonly used to communicate with aerial robots have poor penetration into water [140], and reliable communication is difficult subsurface.

The robot weighs slightly more than 100 g. This mass includes batteries, control electronics, a waterproof fuselage and large lifting surfaces. The addition of servo driven control surfaces to the tail and a propeller for aerial propulsion would not significantly increase the take-off weight of the vehicle, and would allow sustained flight from water. The water jet thruster (40% of robot mass) is twice as heavy as is required by design standards due to test safety concerns (Sect. 6.1), and its weight can be reduced without affecting performance. The power consumption of the valve actuation and wing deployment is small (1.3 mAh/launch) and will have little effect on the provision of battery power for flight. The typical range of other MAVs at the 100 g scale is around 1 mile, which is also the required range for a water sampling vehicle given in [141].

6.8.1 Conclusion

In this chapter the design of a reusable water jet thruster which produces repeatable, predictable thrust has been presented. The water jet system has been intentionally designed to lend itself to modification, and the air and gas tanks can be easily refabricated with different dimensions, so power can be tailored to the needs of a mission, and a slightly larger gas tank would allow the robot to operate in rougher water, or carry larger payloads. The SMA valve actuation system has performed over 100 actuations since fabrication, without any noticed loss in performance. The robot is also a lightweight, self-actuated compressed gas storage system, with a 90 bar release capacity (limited by the SMA yield strength), a system which has many other potential applications outside of propulsion, such as inflatable structure deployment or buoyancy control.

The thruster was then developed into a novel, water jet propelled aquatic jumpglider which demonstrated powerful aquatic jumps, and was the first demonstration of jumpgliding from beneath the water surface, though many more have since followed. A theoretical model has been developed which will allow mission-specific design of future robots. The next chapter will combine this system with systems for powered flight (Chap. 7), while a later chapter (Chap. 9) will describe an extension of this system to allow multiple, repeated aquatic escape.

Chapter 7
Airframe Design for Plunge Diving

Check for updates

7.1 Introduction

In this chapter the design of a plunge diving AquaMAV is detailed. This enhanced AquaMAV prototype is capable of propelled flight, wing retraction for diving into water and jet propelled aquatic escape. The selection process for key components is detailed, as well as the specific attributes necessary for aerial-aquatic locomotion. The AquaMAV includes some limited automation, to allow the robot to self launch when in water, where radio communication is challenging.

The aerial-aquatic cycle of the AquaMAV is condensed in Fig. 7.1. The 160 g robot flies with its wings deployed and its electric propeller provides thrust. The dive is remotely initiated through the folding of the wings. The propeller is turned off, which folds the blades. As the robot hits the water, the wings and propeller are fully folded back. The reduced frontal area and speed allows the robot to reach depths of up to 3 m below the surface. The AquaMAV then slowly and passively floats back to the surface in a nose-up position. The aquatic escape mechanism presented in Sect. 6.4 propels the robot to flight speeds. The propeller is turned on and level flight regained.

7.2 The AquaMAV Airframe

Many morphing aircraft have been demonstrated at the 1 m wingspan scale [142], most commonly to allow UAVs to fit into smaller containers for transport or launch [143, 144], but also to expand the flight envelope of the vehicle in the air. A miniature flying robot able to dynamically change its wing sweep is able to operate in a broader design envelope, and enhance its agility in complex requirements [91], without compromising efficiency in level flight. Sweeping wings backward has also been shown to allow a vehicle to improve its efficiency at higher speeds by reducing parasitic drag [145]. In nature, wing morphing is a near ubiquitous feature of flying

R. Zufferey et al., *Between Sea and Sky: Aerial Aquatic Locomotion in Miniature Robots*, Biosystems & Biorobotics 29, https://doi.org/10.1007/978-3-030-89575-4_7

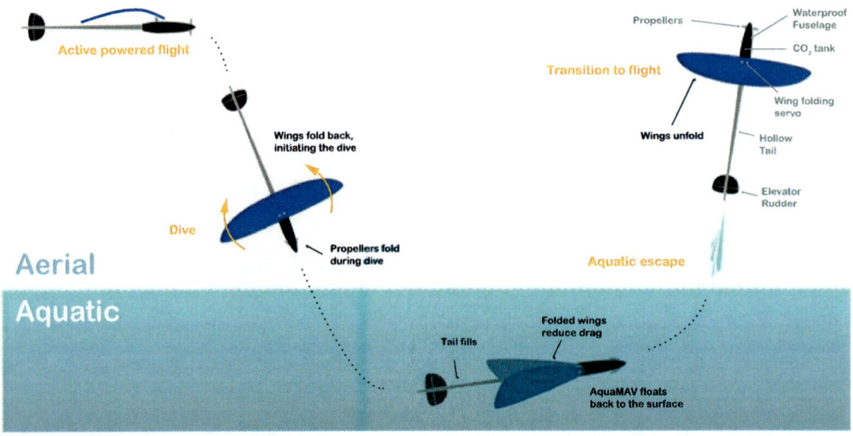

Fig. 7.1 Envisaged multi-modal AquaMAV operation cycle, involving: flight, plunge-dive manoeuvre, water impact and entry, underwater operation, water exit and transition back to flight

animals, and has been of interest to many biologically inspired robots, particularly those with multimodal mobility. Reconfigurable wings are used in jumpgliding robots to decrease drag during ascent [146], in aerial terrestrial robots to enhance mobility on the ground [95], or to mitigate hydrodynamic loads in aerial-aquatic robots [25, 114].

In Chap. 3, it was argued that the plunge diving of predatory seabirds offer an effective solution to water entry for small robots. Gannets reconfigure their wings continuously as they dive, sweeping partially back at the start of a dive but, only fully retracting at the point of impact with water. However, this is principally to allow the bird to accurately track prey as they descend toward the surface [147], and is a level of accuracy not necessarily requisite in a field robot. As an AquaMAV does not necessarily need the aerodynamic control nor the ability to self propel with its wings, the same behaviour can be achieved with less complex, more robust mechanics. The gannet's head and beak are also adapted to mitigate impact loads with the water [148], protecting their neck and face from damage. However, the flexibility of the bird's neck is a consequence of its need to visually track prey, and it again may not be necessary to directly mimic the bird's head structure in order to impact with the water.

The AquaMAV is a small aircraft designed to fly to a target location where it can dive into the water and gather data, before retaking flight and returning to submit samples for analysis. The prototype presented in this chapter is a 201.3 g fixed wing aircraft (Table 7.1), building upon the previous 100 g prototype detailed in Sect. 6.4, which demonstrated take-off from beneath the water. Once in the air, this vehicle could deploy wings and transition to gliding flight, but was unpowered (Fig. 6.11). The new prototype retains the ability to launch from water, again using the jet system outlined in Sect. 6.1, with the gas tank increased in size by a factor of 1.7, and the water tank resized accordingly (Sect. 6.2.1). The mass of the new jet is not significantly

Table 7.1 Component mass breakdown for the AquaMAV

Component	Mass (g)	/Total (%)
Wings	41.0	20
Fuselage	27.1	13
Propulsion	23.5	12
Water Jet	37.0	18
Empennage	8.5	3
Control Servos	15.5	8
Electronics	22.7	11
Batteries	26	13
Total	**201.3**	**100**

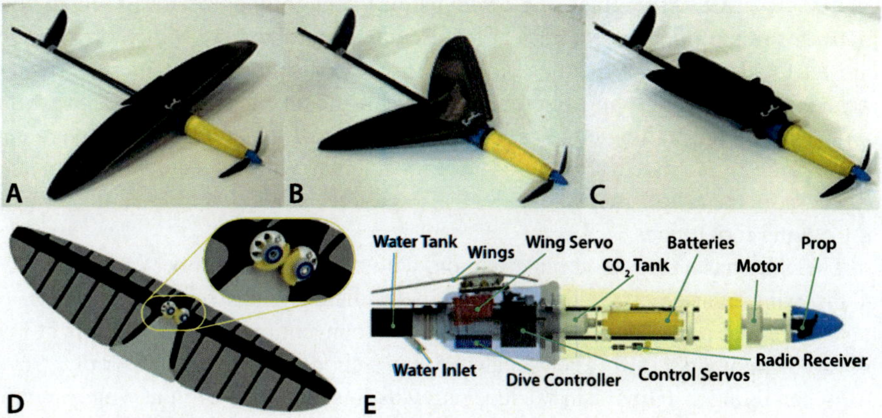

Fig. 7.2 The AquaMAV Airframe: (**A–C**) Fabricated prototype, shown with wings open (**A**) and swept 45° (**B**) and 90° (**C**) backward. (**D**) CAD rendering of the wings, showing the layup pattern and gear linkage. (**E**) CAD illustration of fuselage, with internal components labelled

increased due to a reduction in wall thicknesses, which were previously twice that required by European standards [133]. These performance increases have allowed the provision of propeller propulsion, aerodynamic control and a more developed folding wing design which is more efficient in flight, and with which the wings can be swept backwards up to 90° (Fig. 7.2A–C). This allows the vehicle to dive directly into water without damage, or partially sweep its wings to change its flight behaviour in air.

The new AquaMAV prototype has a conventional fixed-wing layout with a tractor propeller configuration, and a V-tail with dual elevons for control. The V-tail's 45° dihedral angle provides sufficient control in roll, pitch and yaw, whilst minimising the number of actuators required, and the positive dihedral also provides some additional roll stability. The robot is provided with sufficient battery for 14 min of flight, which corresponds to approximately 4 km of sample return range at its 10 ms^{-1} flight speed

(Table 7.2). The keel of the robot is formed by the aluminium CO_2 tank and hollow carbon tube which make up the jet. The robot's control electronics and actuators are all contained in a waterproof kevlar fuselage, mounted to the aluminium gas tank (Fig. 7.2E).

The AquaMAV's main wing is separated into three segments, of which two can sweep backward, while the third is fixed directly to the fuselage, but is compliant enough to allow the other wing segments to pass beneath it (Fig. 7.2D). The two rotating segments pivot on sealed bearings attached to the wing at the quarter chord line. The wing chord line is fixed at a 5° angle of attack relative to the rest of the aircraft, and so as the wings sweep backward, the two segments pass over the fuselage, deforming and reducing their angle of attack relative to the aircraft. The fully open wing has a Zimmerman planform, which has elliptical leading and trailing edges, joined at the quarter chord line, and is a common choice for fixed wing vehicles at the 10^{-1} kg scale. While this wing aerodynamically slightly outperforms rectangular platforms of the same aspect ratio at low Reynolds numbers [149], it was primarily chosen for its suitability to the folding design. Most importantly, the tapering of the wing towards the wing tip reduces the deformation needed to fold the wing completely as the trailing edge slides over the fuselage, meaning a stiffer wing can be more easily actuated. Compared to a rectangular wing of equivalent area and aspect ratio, the elliptic wing shape will also produce smaller aerodynamic moment at the wing root hinges.

The wing has a cambered plate section, with a maximum camber of 5%, located at the wing quarter chord line. Such thin cambered plates have been shown to have increased performance when compared to conventional thick aerofoils at low Reynolds numbers [150]. The AquaMAV has a chord Reynolds number of 7.96×10^5 when flying at $10\,\mathrm{ms}^{-1}$ in sea level air, which is the upper limit at which significant performance benefits are seen, but the use of a cambered section has particular advantages to the folding design: When rotated 90°, a thin cambered wing has a very small frontal area, and so the wing drag when swept fully backward will be greatly reduced compared to a thick wing. Furthermore, a thick wing would have significant buoyancy if watertight, which would inhibit effective locomotion underwater, and if water-permeable the wing would temporarily retain liquid during take off and slow the robot down.

The upper surface of the wing is made from 230 gm^{-2} woven prepreg carbon fibre. This is reinforced by a second layer of woven fabric composed of a spanwise strip along the wing quarter chord line, reinforcing the point at which the wings are attached to the hinges and a location with high cross-section curvature, where it adds the most to the wing's bending stiffness. Six perpendicular strips are also cut into this layer to add to the trailing edge stiffness, with additional reinforcement on the inside edge of the wing, to minimise the risk of cracking or fraying of the edges as the wings slide over each other during folding (Fig. 7.2D).

The two rotating segments are geared together to ensure symmetry, and are driven by a single 6.5 g servo (MKS DS65K, see Sect. 13). The rotating wing segments are attached with screws to the acetal gears, via a 3D printed piece which the matches the cambered wing surface to the flat gear. A 6 mm diameter aluminium rod runs

Table 7.2 Summary of key AquaMAV aerodynamic parameters

Open Wing Span	592 mm
Open Wing Root Chord	120 mm
Open Wing Area	0.056 m^2
Open Wing Aspect Ratio	6.35
Tail Area	0.010 m^2
Tail Aspect Ratio	3.57
Flight Speed	10 m/s
Flight Reynolds Number	79,6000
Battery Capacity	650 mAh
Flying Time	14 min

through each wing into a bearing which is bonded in turn to the fuselage, which is reinforced a the wing attachment point by a carbon fibre plate. Two additional bearings are bonded to centre segment, through which it is bolted to the main wings, allowing the wings to rotate beneath it, while keeping the shafts partially supported at the upper surface.

The internal components of the vehicle can be seen in Fig. 7.2E, showing the servos used for wing deployment and tail actuation, the motor for aerial propulsion and the custom aluminium gas tank used to store high pressure CO_2 for underwater take off (Table 7.2). This is all contained within a streamlined kevlar fuselage which can be separated by a screw connection to access internal components without compromising the waterproofing (this connection is located just ahead of the wing leading edge, joining the rear, blue coloured section to the forward yellow section in Fig. 7.2E). While the vehicle fuselage was designed to be as slender as possible to minimise drag and impact forces, the cross section was primarily determined by the need to accommodate the internal components while allowing easy disassembly. The nose shape is a revolved NACA 0010 airfoil, chosen to accommodate the propulsion motor, batteries and desired propeller.

One of the most challenging practical aspects of the AquaMAV's design is waterproofing, to ensure appropriate buoyancy and to protect components from water damage. To actuate the external moving parts with the internal motors some custom built sealed bearings were used, by laser welding stainless steel shields onto rubber sealed bearings and impregnating them with water repellent grease. The motor was also sealed from the remainder of the fuselage so that internal pressure limited any potential water ingress. Control lines to the tail elevons were run through 0.6 mm teflon tubes which were also impregnated with grease. Finally, the servos were disassembled and their casings filled with grease and flight control circuit boards were coated with an acrylic laquer, so that in the event of seals failing, damage to internal systems was mitigated.

7.2.1 Control

The AquaMAV is flown remotely over radio when in flight, at which point the operator has control over propulsion, elevons, wing folding and actuation of the jet. However, when in the water, the 2.4 GHz radio signal has poor penetration, and the robot may not be able to receive signals telling it to launch itself. To deal with this, a simple level of autonomy has been added, allowing the AquaMAV to self launch using a conductivity sensor placed in its nose, just behind the propeller. The sensor is a simple pair of brass contacts, between which current can flow when the robot is underwater.

When the AquaMAV enters the water, its water tank is empty, and it will float with its nose facing downward. The water tank gradually fills with water through an inlet check valve (Fig. 7.2E) which shifts the centre of buoyancy and makes the AquaMAV float nose up. A small microcontroller detects when the AquaMAV enters the water, at which point it waits until a specified time after this signal stops. At this point the AquaMAV is floating nose up, when the nose sensor will be out of the water, the water tank is full and the jet is ready to fire. In this way an AquaMAV can recover from a dive and relaunch without external input.

7.2.2 Propulsion

To select an appropriate propulsion system for the AquaMAV, the propeller analysis program QPROP [151] was used. QPROP computes the steady-state behaviour of propeller systems using an enhanced blade-element method built on the method of Larrabee [152]. It computes an accurate propeller aerodynamic model by accounting for the propeller's self-induction. The propeller is defined geometrically using a series of chord and twist values along the blade, and aerodynamically by Sect. 2D lift and drag coefficients. Definition files for propellers were taken from publicly available databases [125, 151] or, where information was unavailable, purchased propellers were digitised using a 3D scanning system (see Sect. 10.3 for details).

First, QPROP was tested against experimental data from [125] for propellers at appropriate Reynolds numbers, to validate predictions. The experimental data is expressed in terms of advance ratio, J a dimensionless measure of the ratio of blade tip speed to freestream velocity, v:

$$J = v/D\omega \tag{7.1}$$

where D is the propeller diameter and ω is its rotation speed. Thrust, T is also expressed dimensionlessly as

$$C_T = T/\rho_a D^4 \omega^2 \tag{7.2}$$

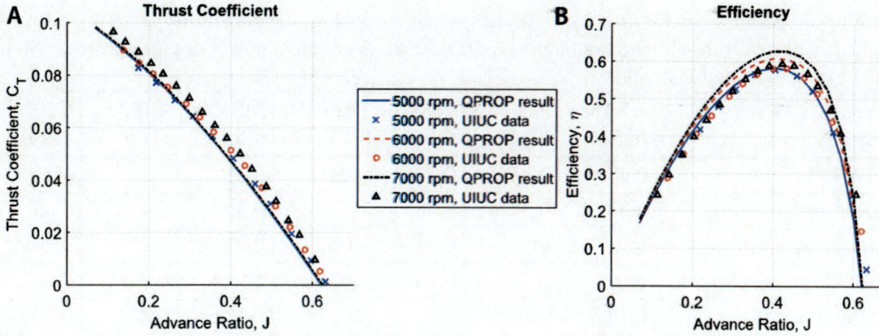

Fig. 7.3 QPROP validation by comparison of predictions with experimental data from [125] for an APC 203 × 97, a 203 mm diameter 97 mm fixed pitch commercial propeller. (**A**) Thrust Coefficient against advance ratio. (**B**) Efficiency against advance ratio

where ρ_a is the density of air. Propeller efficiency is defined by the ratio of propulsive power out (Tv) to shaft power in (shaft torque, Q multiplied by angular speed, ω),

$$\eta_P = Tv/Q\omega \tag{7.3}$$

Figure 7.3A shows measured and predicted efficiency curves for an APC 203 × 97 propeller. The predictions show good agreement with experiment, with some discrepancies likely due to simplifications in the propeller definition.

To develop an efficient propulsion system, characterising propeller behaviour alone is insufficient, and a high performing propulsion system must be well matched to the drive motor. To do this, a first-order motor characteristic is used, given by:

$$Q = \left((V - \frac{\omega}{K_v})\frac{1}{R} - I_0 \right) \frac{1}{K_v} \tag{7.4}$$

where the relation between the shaft torque and rotational speed is given by voltage, internal resistance (R), no-load current (I_0) and a speed-voltage coefficient (K_v), the latter three quantities being characteristic constants of the motor. Motor efficiency is the ratio of shaft power to electrical power:

$$\eta_M = Q\omega/VI \tag{7.5}$$

where V and I are the input voltage and current. Total system efficiency is then:

$$\eta_T = \eta_P\eta_M = Tv/VI \tag{7.6}$$

The motor model can then be used in conjunction with propeller geometry files to evaluate the most appropriate propulsion system. This was done by evaluating the total motor-propeller efficiency at a freestream speed of 10 ms^{-1}, where the motor is

Table 7.3 Motor-Propeller total efficiencies when producing 25 gs of thrust at 10 ms^{-1}. An ⊘ indicates that the motor could not operate, either due to insufficient power or current limits. Motor masses are indicated in brackets

Propeller Motor	AX1306 (6.5 g)	Park180 (8 g)	18MOR (10 g)	Park250 (13 g)	MS1704 (15.5 g)	C1826 (18 g)	C2028 (22 g)
GWS 127 × 76	0.29	0.32	0.32	0.33	0.31	0.36	⊘
Eflite 130 × 70	0.33	0.36	0.38	0.39	0.36	0.41	0.43
CAM 152 × 76	⊘	⊘	0.23	0.24	0.21	0.25	0.26
APC 178 × 89	⊘	⊘	0.23	0.24	⊘	0.25	0.26
APC 203 × 97	0.14	0.14	0.2	0.22	0.18	0.22	0.23

Table 7.4 Motor-Propeller static thrusts in Newtons when driven at 7.4 V (maximum throttle). An ⊘ indicates that the motor could not operate, either due to insufficient power or current limits

Propeller Motor	AX1306 (6.5 g)	Park180 (8 g)	18MOR (10 g)	Park250 (13 g)	MS1704 (15.5 g)	C1826 (18 g)	C2028 (22 g)
GWS 127 × 76	1.17	1.18	1.4	1.87	1.7	2.2	0.81
Eflite 130 × 70	1.16	1.16	1.44	1.94	1.73	2.29	0.85
CAM 152 × 76	1.41	⊘	1.96	2.73	⊘	3.18	1.25
APC 178 × 89	1.67	⊘	2.52	⊘	⊘	⊘	1.73
APC 203 × 97	1.81	⊘	⊘	⊘	⊘	⊘	2.53

producing 25 g of thrust (based on a 200 g flying weight and a conservative lift to drag ratio estimate of 8). In order to ensure that the AquaMAV could climb effectively during takeoff, static thrusts was also evaluated, and a static thrust to weight ratio of >1, or >2 N thrust was targeted. This is tabulated for a variety of commercially available folding propellers and motors in Tables 7.3 and 7.4. The propellers are identified by their manufacturer and a number, for example '152 × 76', referring to a propeller with a 152mm diameter and a 76mm blade pitch. Motors are identified by manufacturer's model number. A 'Park250' motor and 'Eflite 130 × 70' propeller were selected as a compromise between efficiency, motor mass and static thrust. Slightly higher performance could be achieved with a C1826 motor, but not only is this motor heavier, but its larger current draw necessitates heavier electronics.

The AquaMAV requires three servos to actuate its wings and control surfaces. In order to select the most appropriate component, and to reduce as far as possible the mass of the vehicle, a database was made of servos in the appropriate size range, and performance parameters were developed in terms of torque/weight and power/weight, and the highest performing servo was chosen.

7.3 Wind and Water Tunnel Measurements

Tests were conducted in a closed circuit wind tunnel, with a 0.95×0.95 m working section. Wind speed was measured and controlled using a pitot probe, compensated for changes in ambient conditions. During tests, airspeed was maintained to within 0.5% of demanded velocity.

The vehicle was mounted to a multi-axis force balance (ATI Gamma IP68) via a 0.5 m streamlined steel sting, such that the centre of the fuselage was located 0.45 m above the tunnel floor at 0° angle of attack. The force balance was in turn attached to a custom-built worm drive system, which was used to vary the vehicle angle of attack, α. The 200 step/rev stepper motor used to drive the worm gear gave an accuracy of 7.5×10^{-4} °/step after the 1:150 reduction in the worm drive (Fig. 7.4A).

The force balance has a calibrated range of 100 N force and 10 Nm torque in each axis, with force and torque resolutions of 25 mN and 1.25 mNm respectively. Data from this sensor was sampled at 10 kHz, averaging 1000 samples per recorded reading. During testing, zero readings were recorded for 60 s before and after the tunnel was run and averaged together to account for sensor drift. Aerodynamic forces were also recorded for 60 s, once the tunnel had reached the demanded speed.

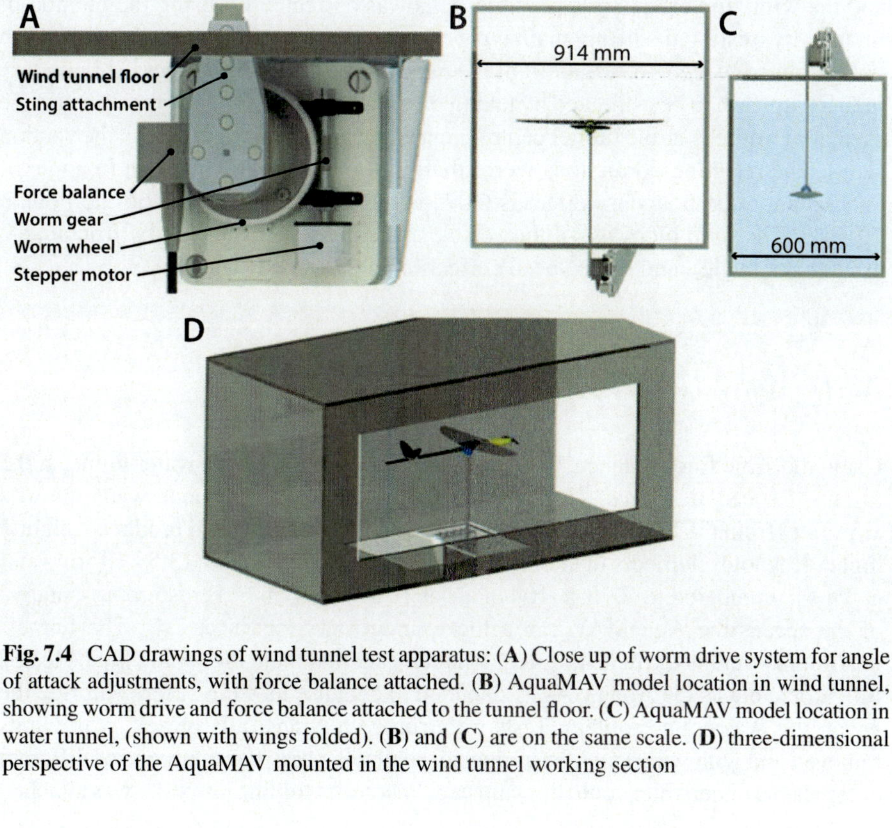

Fig. 7.4 CAD drawings of wind tunnel test apparatus: (**A**) Close up of worm drive system for angle of attack adjustments, with force balance attached. (**B**) AquaMAV model location in wind tunnel, showing worm drive and force balance attached to the tunnel floor. (**C**) AquaMAV model location in water tunnel, (shown with wings folded). (**B**) and (**C**) are on the same scale. (**D**) three-dimensional perspective of the AquaMAV mounted in the wind tunnel working section

The wind tunnel, force balance and worm drive were all operated using LabView, which was also used to control the wing servo, allowing the effect of wing sweep angle to be evaluated. The open wing was tested from −6° to 24° angle of attack, in 2° increments at its design flight speed, $10\,\mathrm{ms}^{-1}$, to elucidate the complete stalling behaviour of the wing. The wing was also tested at different sweep angles, sweeping the wings backward in 10° increments from fully open (0°) to fully closed (90°), also at $10\,\mathrm{ms}^{-1}$. The fuselage was also tested with wings, propeller and empennage removed to evaluate the different drag contributions, and to calculate separate aerodynamic coefficients for the wings and fin, so that any pitch damping effects could be accounted for in trajectory modelling.

It was necessary for the wing to have some compliance in order to fold, as well as being as thin as possible to reduce weight. In order to evaluate the effect of this compliance, a rigid wing with the same planform and camber profile was fabricated from a 2 mm thick aluminium plate, and tested separately, so that a performance comparison could be made.

Once tunnel measurements were collected, the sting was run separately over a large range of angles of attack and tunnel speed. The forces and moments produced by the sting were then subtracted from the model measurements taken at the corresponding speeds and orientations. The maximum wingspan of the vehicle (0.592 m, 73% of tunnel width, (Fig. 7.4B)) meant that some interference between the tunnel walls and the wing tip vortices was possible. This was corrected for using the method of images, by treating the lifting main wing as a pair of trailing vortices whose strength is determined by geometry and lift of the main wing [153]. This allows the boundary induced upwash to be estimated by taking the upwash induced by the resulting image system of vortices at the tunnel centreline as the average upwash across the section [153]. The resulting corrections were small. For example, the increase in angle of attack due to the boundary effect is 0.09° with the AquaMAV at 0° to the flow at $10\,\mathrm{ms}^{-1}$. The solid blockage of the vehicle and sting was low (1.4% by frontal area at 0° attack angle), and so no speed correction was deemed necessary.

7.3.1 Water Tunnel Setup

Using the same force balance, the AquaMAV was also tested in a water flume, at 0.2 to $1\,\mathrm{ms}^{-1}$ speed, in $0.2\,\mathrm{ms}^{-1}$ increments. Taking measurements underwater allowed any effect from the larger hydrodynamic loads to be observed, and produced slightly higher Reynolds numbers than those observed during tests in air (13.5×10^4 in water at $1\,\mathrm{ms}^{-1}$ compared to 7.96×10^4 in air at $10\,\mathrm{ms}^{-1}$). $1\,\mathrm{ms}^{-1}$ is also representative of the speeds the AquaMAV can achieve underwater (see Sect. 7.4). The flume's maximum width is 0.6 m (Fig. 7.4C), which meant that the wings could not be tested fully open, but as the AquaMAV is intended to collapse its wing before entering the water, this was not a problem. Tests were run with wings fully closed, with wings removed and with wings and fins removed, so that the drag contributions of different components underwater could be evaluated. Where the folding propeller was attached

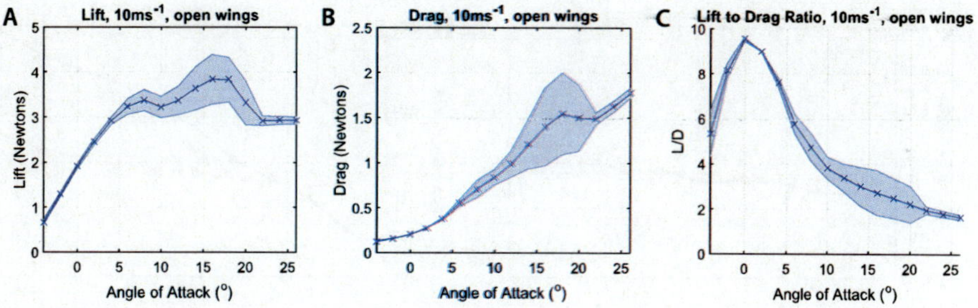

Fig. 7.5 Lift and drag of the AquaMAV airframe in air, as measured in a wind tunnel, with the RMS error of the force measurements indicated by the shaded region. Increased error can be seen as the wing nears stall due to the unsteady flow: (**A**) Lift force in Newtons. (**B**) Drag force in Newtons. (**C**) Lift to Drag Ratio

to the fuselage it was prevented from rotating, although the propeller blades could fold backward freely. This was deemed reasonable as in practice resistance from the motor and friction where the blades are folded flat against the fuselage will limit autorotation.

7.3.2 Results

At its design cruise speed, the AquaMAV performs well, producing sufficient lift to balance its weight (Fig. 7.5A), and in level flight is at its maximum lift to drag ratio (Fig. 7.5C). The vehicle begins to stall at 8°, but full stall is delayed until 18°. However, at lower speeds, this delayed stall is not observed, and the indication is that this is due to wing compliance. This is investigated further in Sect. 7.3.2.2.

Figure 7.6A and B show the forces generated by the fuselage and wing fins. The fins show linear C_L behaviour over the tested range. Fin stall is not observed because of the fins' 45° dihedral angle, which reduces the relative flow normal to each fin and hence the effective angle of attack of fins relative to that of the fuselage. Figure 7.6C and D show the effect of fully retracting the main wing. It can be seen that the drag in level flight is reduced by 62%, and lift production is decreased by 92%.

In order to compute dive trajectories, the lift and drag contributions of the fuselage and fins are extracted separately, shown in Fig. 7.6. For each component, C_L and C_D are calculated from lift, L and drag, D as

$$\begin{pmatrix} C_L \\ C_D \end{pmatrix} = \begin{pmatrix} L \\ D \end{pmatrix} \frac{1}{\frac{1}{2}\rho V^2 A_{ref}} \tag{7.7}$$

where ρ is the fluid density, V the freestream velocity and A_{ref} a reference area. The fully open wing planform area is used as the reference area for all calculated coefficients. The wing centre of pressure is computed from the measured torque on

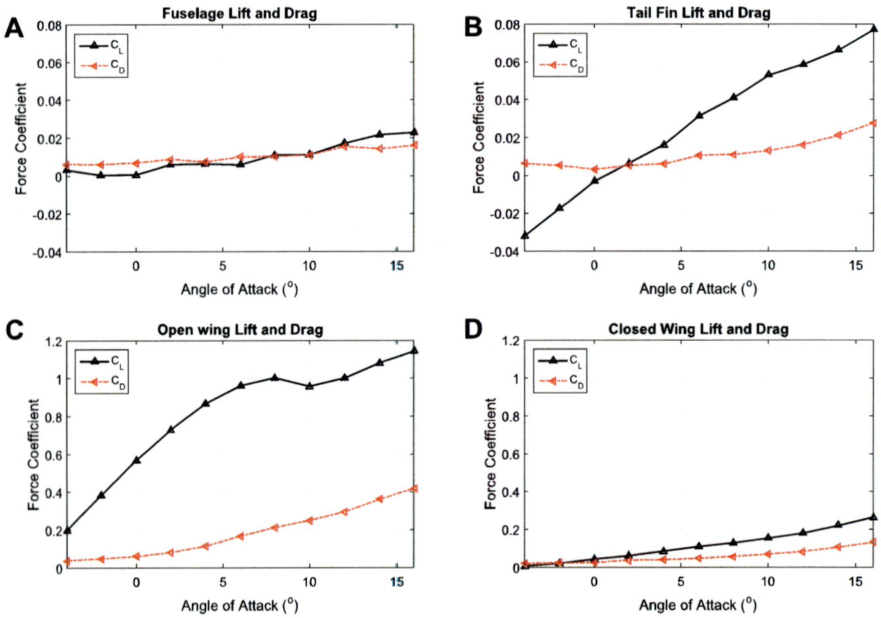

Fig. 7.6 Lift and drag of fuselage and fins: (**A**) Fuselage only, measured with wings and tail removed. (**B**) Tail fins only, with fuselage lift and drag subtracted from measurements. (**C**) Open wing only. (**D**) Fully retracted wing only. Note that **A** and **B** are plotted on a different scale to **C** and **D**

the force sensor, defined as the point of action of the aerodynamic forces about which there is no moment. This point is assumed to lie along the wing chord line, which is at a preset angle of attack, α_0 to the fuselage (5°). By taking moments around the force transducer, the location of the centre can be equated to the measured torque, M:

$$M\hat{\boldsymbol{Z}}_{sensor} = \boldsymbol{R}(\alpha)\mathbf{F} \times \left(\left(S + r\cos(\alpha_0)\right)\hat{\boldsymbol{X}}_{sensor} + r\sin(\alpha_0)\hat{\boldsymbol{Y}}_{sensor}\right) \quad (7.8)$$

where $\hat{\boldsymbol{X}}_{sensor}$, $\hat{\boldsymbol{Y}}_{sensor}$ and $\hat{\boldsymbol{Z}}_{sensor}$ are unit vectors defining the sensor coordinate system, S and r are the distances from the intersection of chord line and sting axis to the sensor origin and centre of pressure respectively (Fig. 7.7A). $\mathbf{F}$ is resultant lift and drag force and $\boldsymbol{R}(\alpha)$ is a rotation matrix about $\hat{\boldsymbol{Z}}_{sensor}$. Given the sensor readings and equipment geometry, Eq. 7.8 can be solved for r at each measurement point.

Figure 7.7B shows the effect of wing retraction on the location of the wing's centre of pressure. The vehicle's centre of gravity (c.g.) is located 205 mm from the vehicle's nose (Fig. 7.7B). When open and at 0° angle of attack, the wing's aerodynamic centre is 212 mm from the nose, slightly behind the c.g. and at 32% of the wing chord, but when the wing is folded, this moves significantly rearward to 313 mm from the nose. This movement of the centre of pressure and the loss in lift means that once the wing

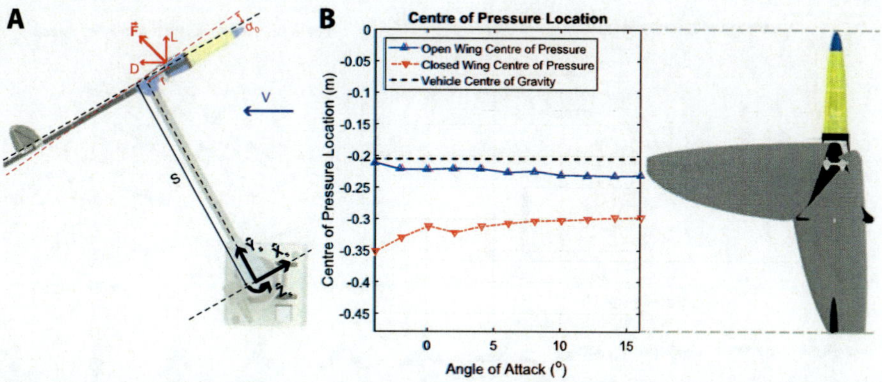

Fig. 7.7 Centre of pressure of the wing and its variation with angle of attack: (**A**) Centre of pressure definition from measured forces and moments at the sensor. (**B**) Measured centre of pressure of the open and closed wing, defined here by the distance from the AquaMAV nose, with the AquaMAV shown on the same scale for illustration

is retracted, the AquaMAV will passively enter a stable dive, the trajectory of which is explored further in Sect. 8.1.2.1.

7.3.2.1 Partial Wing Sweep

With the wings fully open, the vehicle will only be able to increase its level flight velocity beyond the design cruise speed by operating at slightly negative trim angles using the tail control surfaces, which is aerodynamically inefficient. Since the wing folding mechanism allows for the wings to be operated at partial sweep angles, the possibility of sweeping the wings backward to enable the AquaMAV to sprint for short periods was investigated. Figure 7.8 shows the effect of sweep angle on lift production (Fig. 7.8A) and efficiency (Fig. 7.8B).

As the wings sweep backward, they slide over one another and the fuselage, deforming in order to do so, until they reach a maximum angle of 90°, where a hard stop is reached (Fig. 7.8C). In Fig. 7.8A the effect of the wings deforming as they sweep back can be observed, in the rapid drop off in lift over angles of 60–90°, where the wings are forced upward by the fuselage, decreasing their angle of attack relative to the freestream. However, at 10° the wing shows a slight increase in lift coefficient at $\alpha = 0$ (C_{L0}) and aerodynamic efficiency. This is due to the wing trailing edge deflecting downward as it passes under the centre segment, increasing its camber. At 30°, the wing trailing edge reaches the fuselage, is deflected upward and lift production begins to decrease. If the vehicle were to sweep its wings back by 70°, C_{L0} would reduce to 0.15 which would allow the vehicle to maintain a level flight speed of 19 ms^{-1}. This would incur a significant drag penalty (Fig. 7.8B), and require high thrust, but there is nonetheless the potential for wing sweep to be used in a brief sprint in spite of the inefficiency. Sweeping the wings backward also allows the robot

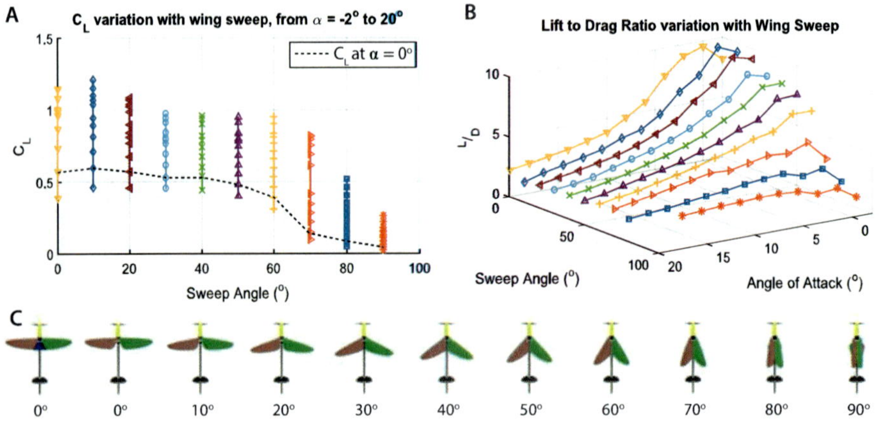

Fig. 7.8 The effect of partially sweeping the wings on wing aerodynamics: (**A**) The range of lift coefficients produced at angles of attack between −2° and 20° in 2° increments, plotted for different wing sweep angles. (**B**) Lift to drag ratio. (**C**) Wing sweep sequence with the centre segment removed, showing the interference between segments as they fold. Image at 0° sweep is repeated, with the first image showing the centre segment

to shift its aerodynamic centre rearward and increase its static margin, altering the handling qualities of the vehicle.

7.3.2.2 Wing Performance at Different Speeds

To investigate the impact of the segmented folding mechanism on the wing design, the fully open main wing was tested at different speeds and compared to a single-piece rigid aluminium wing. The lift and drag measured during these tests are shown in Fig. 7.9A and B respectively. From Fig. 7.9A it can be seen that the rigid wing produces slightly more lift than the folding wing at small angles of attack, which suggests that some warping of the carbon fibre wing occurred after it was demoulded, or a slight manufacturing error in the preset angle of the wing chord relative the AquaMAV fuselage exists. It can also be seen that the carbon wing stalls earlier than the rigid version at low speeds. This could be the result of the protrusions on the wing upper surface which are necessary for the folding mechanism, as well as the discontinuities between the segments, both of which will retard flow over the upper surface, and encourage separation. However it could also be the result of an interaction between the fuselage wake and the wing, as the rigid wing was not tested attached to the AquaMAV body.

While the rigid wing exhibits the same lift behaviour at the three tested speeds, the early stall in the compliant wing is only observed at low speeds. At 8 ms^{-1}, the lift curve shows an inflection near the 6 ms^{-1} stall point, but lift increases beyond. At 10 ms^{-1}, further lift increases are seen beyond 6° attack angle. This disparity at different speeds may too be the result of wake interaction, or the presence of a laminar

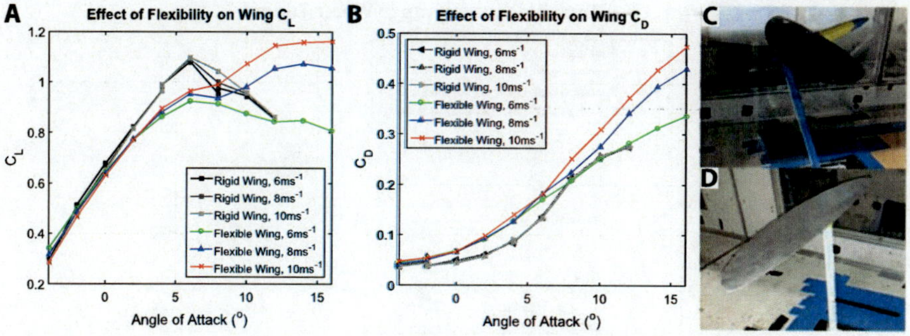

Fig. 7.9 The effect of the folding design on the open wing's aerodynamic efficiency: (**A**) Lift curves for the main wing at 6, 8 and 10 ms^{-1} compared to a rigid aluminium wing at the same speeds. (**B**) Drag coefficients for the rigid and folding wing. The fuselage drag, measured separately is subtracted from the folding wing measurements, but the rigid wing still exhibits lower drag, suggesting that the flow around the fuselage interferes with the wing detrimentally. (**C**) Image of the AquaMAV main wing in the wind tunnel, (**D**) Image of the rigid aluminium wing in the wind tunnel

separation bubble [154]. During the tests, deformation of the wing was observed, both a deflection of the wing trailing edge and a slight increase in dihedral. The wing's compliance, with the wing deforming upward under the increased aerodynamic loads at higher speeds may also contribute to the delayed stall. However, in order to fully determine origin of the disparity it would be necessary to measure either the wing deformation or the flow geometry directly. Nonetheless, the delayed stall will allow the vehicle to be more robust to perturbations in flight.

It was necessary to test the AquaMAV wing attached to the fuselage due to its folding mechanism (Fig. 7.9C). However, as it is being compared with the plain aluminium wing (Fig. 7.9D), the lift and drag of the fuselage as measured without the wing attached is subtracted from the data shown in Fig. 7.9. Examining Fig. 7.9B it can be seen that the rigid wing shows lower drag prior to stall, indicating that the protrusions on the upper surface and gear mechanism on the lower surface, as well as interference with the wing flow from the fuselage, all act to increase folding wing drag relative to the rigid planform.

7.3.2.3 Sources of Drag

In Fig. 7.10 the hydrodynamic drag acting on the vehicle at 0° angle of attack is plotted against speed in water. Tests were repeated with the vehicle wings, fins and propeller removed, to show the different contributions (due to the flume's 0.5 m width, the wings were only tested fully folded). For each curve a quadratic polynomial relating drag, D, to velocity, v (with the form $D = Kv^2$, where K is a constant) is fitted by regression, and a drag coefficient of the form in Eq. 7.7 was extracted. The difference in drag between, for example, the fuselage with propeller attached and fuselage without allowed us to extract the drag contributions of different components.

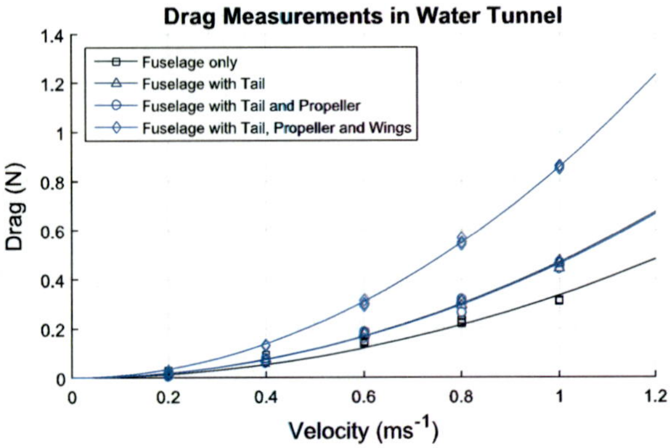

Fig. 7.10 Drag measurements at 0° angle of attack in a water tunnel, with a quadratic curve fitted to extract C_D

Table 7.5 Comparison between measured drag contributions of AquaMAV components in air and water at 0° angle of attack

	Water		Air	
Component	C_D	/Total (%)	C_D	/Total (%)
Fuselage	0.0121	39	0.0069	17
Fins	0.0047	15	0.0031	7
Propeller	0.0002	1	0.0012	3
Wings	0.0142	45	0.0299	73
Total	**0.0312**	**100**	**0.0410**	**100**

Reasonable agreement between tests in air and water was found, although the need to disassemble the AquaMAV between tests may have resulted in some inconsistencies. Notably, the measured drag coefficient of the wing underwater is half that which was measured in air. This is most likely due to bending of the wing. Given the folded wing's C_{L0} of 0.044 (Fig. 7.6D), the wing would produce 0.15 N of force in air at 10 ms^{-1}, but the same lift coefficient in water flowing at 1 ms^{-1} would produce 1.21 N of force.

7.4 Performance Assessment

In the preceding chapters systems which enable aerial-aquatic locomotion have been presented and integrated into a robot, the AquaMAV (Fig. 7.11). In this section the impact of hybridising locomotion on single mode mobility is examined and the costs and benefits of multimodal locomotion are quantified, by combining QPROP simulations of the propulsion system with the aerodynamic and hydrodynamic data

Fig. 7.11 The AquaMAV, with open and folded wings

from wind and water tunnel testing. In order to assess the AquaMAV, cost of transport (COT) is used as a metric, a dimensionless measure of the energy required to travel one meter, inversely proportional to a vehicle's maximum range. It is defined here as

$$\mathrm{COT} = \frac{P}{mgv} \tag{7.9}$$

where P is the power input required for locomotion, v is velocity, m is mass and g is gravitational acceleration. If propulsive efficiency is ignored, $P = Dv$ (where D is drag) and

$$\mathrm{COT} = \frac{D}{mg} \tag{7.10}$$

which is the reciprocal of lift to drag ratio for a flying vehicle.

7.4.1 Aerial Locomotion

The most immediately obvious consequence of adapting the AquaMAV for aerial-aquatic locomotion is the provision of additional components which are unused when swimming or flying alone. The prototype presented weighs 212 g when the gearbox presented in Chap. 10 is included. This weight is broken down in Fig. 7.12, showing the portion of mass which is unused in flight or unused underwater. For effective multimodal locomotion, this redundant mass should be kept as low as possible in both modes. This has been done in the AquaMAV by using the water tank of the jet thruster to form the robot's tailboom, and by packing all the electromechanical components as close together as possible, so that the need waterproof housing does not make the robot significantly heavier than does the normal need for aerodynamic streamlining in a flying vehicle. The same analysis is performed on the SailMAV, a sailing-flying robot which does not entirely submerge. Comparison between those two platforms can be found in Sect. 11.9.1.

In Fig. 7.13A COT is plotted across the range of achievable flight speeds. The lower limit set by the maximum lift coefficient before stall, achieved at an 8° angle of attack

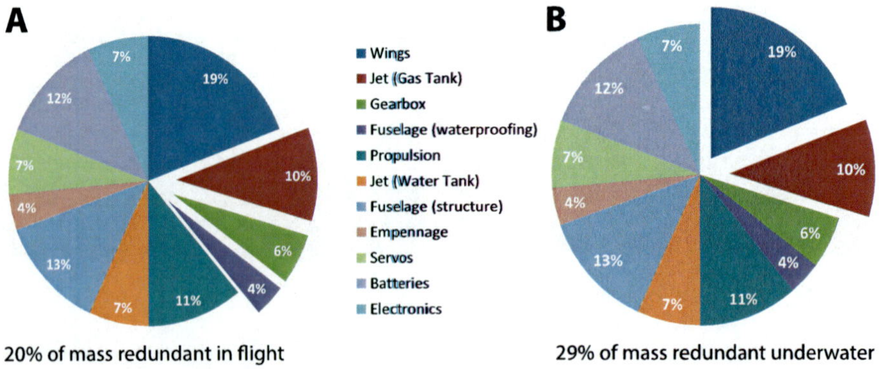

Fig. 7.12 The mass cost of aerial-aquatic locomotion: (**A**) 20% of the AquaMAV mass is unnecessary for flight. (**B**) 29% of the AquaMAV mass is unnecessary for movement underwater

(Fig. 7.5A). The maximum speed is limited by propulsion, top speed being that at which the propeller cannot produce sufficient thrust to balance drag at equilibrium trim. Lift to drag ratio is maximum at 0° angle of attack (Fig. 7.5C), and since at this angle the AquaMAV must fly at $10\,\mathrm{ms}^{-1}$ to balance its weight, this speed is also where COT is lowest.

The vehicle is designed to be near neutrally buoyant, so the presence of additional mass is far less significant for underwater locomotion. However, weight is key in flight due to the need to create sufficient aerodynamic lift. In flight, the AquaMAV must carry the water jet system, the underwater gearbox and waterproofing, which constitute an unused mass of 42 g (20% of total). Assuming that the lift and drag coefficients did not change significantly (reasonable as the flight Reynolds number does not change significantly), if this weight were removed the AquaMAV would require 80% of its current thrust (lift to drag ratio being unchanged), and could travel at 89% of its current flight speed, which would reduce the power required for flight to 72%.

This power reduction would mean a commensurate reduction in COT if propulsive efficiency is ignored. However, the change in speed would also mean that the operating point of the propulsion system changes, and so the COT in Fig. 7.13A shows an increase in energy costs as well as a shift to a lower speed. The additional friction created by the aquatic gearbox is more immediately detrimental to cost of transport than the added weight of the additional systems, particularly given that the hypothetical lighter aircraft's COT does not account for the fact modifications to the propulsion system could increase efficiency.

Overall, it is reasonable to suggest that the AquaMAV is not significantly hindered by the additional subsystems, in terms of locomotion efficiency. The more significant detrimental impact of multimodality is to its handling qualities. In flight, the manoeuvrability of the AquaMAV is reduced compared to another fixed wing vehicle of comparable size; the folding wing design prevents the provision of ailerons on the wings, and as a consequence roll authority is diminished. Secondly, the wings have

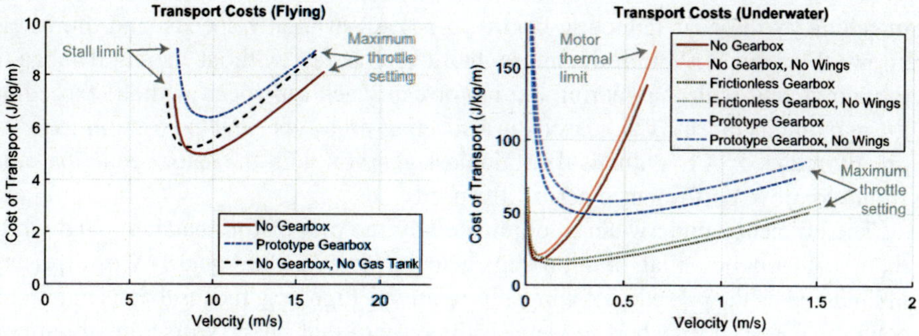

Fig. 7.13 The AquaMAV's cost of transport in air and water, produced by combining QPROP simulations with tunnel test data. Annotations indicate the physical limitations on locomotion speed. (**A**) Flight cost of transport, showing the range of speeds and energetic efficiencies of the AquaMAV's flight, and the cost of the additional gearbox friction if the prototype gearbox described in Sect. 10 is included. The change in performance if the CO_2 tank and other unused components are removed is also shown, which reduces mass. (**B**) Underwater cost of transport, showing the effect the gearbox as well as the effect of removing the AquaMAV wings, which reduces both mass and drag

no dihedral and are not mounted significantly above the centre of gravity (Fig. 7.11), the latter being necessary to reduce the frontal area of the vehicle. Were positive dihedral to be added, it would significantly restrict the ability of the wing segments to fold over one another and increase the frontal profile of the folded wings, and so was avoided. The result of this is an increased tendency for the vehicle to sideslip. This does not prevent the vehicle from flying stably, but can make it challenging for a human operator in certain conditions.

7.4.2 Aquatic Locomotion

In Chap. 10 the use of an aerial propulsion system underwater was examined, and a gearbox presented. Using the drag data gathered in wind and water tunnel testing, how efficiently the AquaMAV is able to move beneath the water can be calculated, considering a steady-state, where propulsive thrust is equal to the drag. Underwater, the folded wings constitute 45% of the total drag (Fig. 7.10, Table 7.5). If the same propulsive efficiency could be obtained, this would give a commensurate reduction in COT, but including the effect on propulsion efficiency reduces the net gain (Fig. 7.13B).

Whilst in flight, the gearbox friction is detrimental across all speeds, as it does not change the gearing between motor and propeller and only adds resistance. However, when operated underwater the change in gearing means that the AquaMAV is able to move significantly faster. The ungeared motor is more efficient at low speeds due to the lack of gearbox friction, but performance rapidly worsens at higher speeds, and quickly reaches the motor's maximum current (thermal limit). The improved matching by the gear reduction begins to pay off at higher speeds, and ultimately allows the AquaMAV to move more than twice as fast without risking damage to the motor, and with the gearing the motor can reach top speed without exceeding the maximum current. The gearbox is also a first prototype, and the friction could be greatly reduced. A hypothetical frictionless gearbox with the same reduction ratio would improve performance across all speeds.

The efficiency underwater is dominated by the propulsion matching, and while there is obvious potential for improvement to the gearbox, the AquaMAV is surprising unhindered by the presence of wings. Its relatively high drag means that its minimum achievable cost of transport in water is three times that in air. Worse, the minimum COT speed is very low ($<0.1\,\mathrm{ms^{-1}}$), and so it will take the robot a great deal of time to travel longer distances, and over long periods of time the small current draws for the control electronics will consume more battery capacity and reduce range. In flight the AquaMAV travels quickly, and these draws are insignificant relative to the motor power. The gearing therefore is of additional benefit.

This analysis assumes that the AquaMAV can float levelly underwater. This is easily achievable with small changes to the fuselage shape, but the current design deliberately makes the vehicle float nose up, ready for launch. In the current configuration, it could potentially use its propeller to achieve greater depth than was estimated for a passive dive, although it would only be able to do so stably for 10–20 s before its water tank filled and buoyancy oriented it nose up. Ideally this would include some sensing of freestream speed, as at the initial high dive speeds driving the propeller would simply create more drag.

7.4.3 Summary of Performance

The AquaMAV moves more efficiently in air than in water, in contravention of the broad trends in animal locomotion (Fig. 7.14). This is in essence a consequence of the AquaMAV's design intent. The robot was intended to collect and return a water sample from the air, and (colloquially) is a 'submersible aircraft' as opposed to a 'flying submarine'. The tail fin size is larger than necessary for stability underwater, even when retracted its wings contribute significant drag, and the vehicle is more slender than is hydrodynamically optimal. Its performance suffers as a consequence. But even if an AquaMAV were designed to primarily move underwater, flight naturally takes design primacy, as it is more restrictive; stall limits mean that there is minimum as well as maximum speed, and the need for lift means a minimum wing area required to fly well without excessive power requirements. This being said, it is worth noting that the primary reason for equipping a vehicle with multiple mobility modes is not necessarily locomotion efficiency, but more often is access and versatility; while the wing folding may impact handling qualities, and the fuselage has excessive drag underwater, this design allows the robot to enter the water reliably with minimal external control, and retake flight in spite of surface motion. These benefits are just less readily quantified.

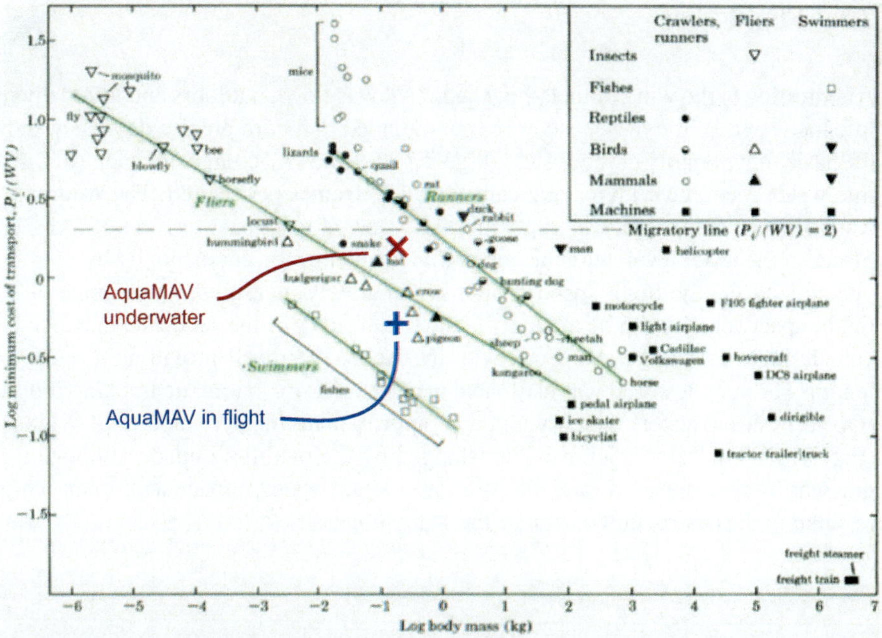

Fig. 7.14 The AquaMAV's lowest achievable cost of transport in air and water in context, showing transport costs for animals and machines (adapted from [155])

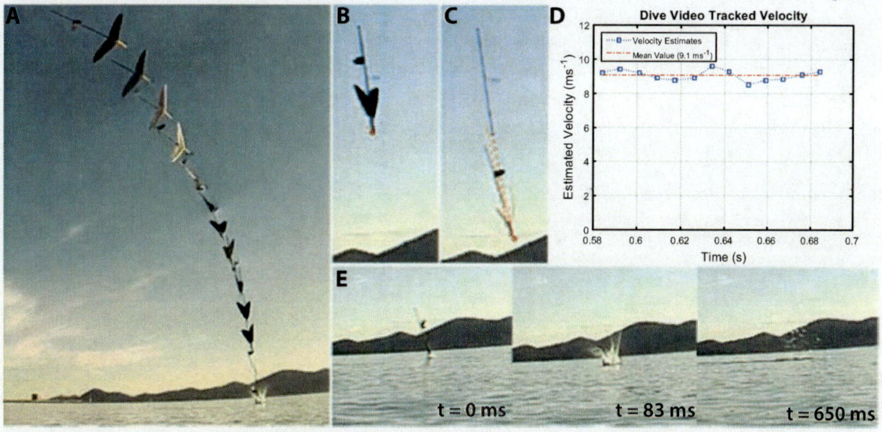

Fig. 7.15 Preliminary flight test of the AquaMAV outdoors. (**A**) Composite image of a dive into water, with frames 83 ms apart. The impact velocity was estimated from the later video frames: (**B**) Calibration frame using vehicle length as reference. (**C**) final frame showing tracked points. (**D**) Velocity estimate from tracked points. (**E**) Image of the AquaMAV at the point of impact with the water. (**F**) Image of the AquaMAV fully immersed after a dive

7.5 Flight Test

After testing in the wind tunnel, the AquaMAV was flown outdoors and tested diving into the water. While these tests were not intended to record precise data, they show the basic functionality of the system. Figure 7.15A shows a composite image of a dive into water, recorded with a static camera at 120 frames per second. The AquaMAV was launched from an elevated position and steered to impact in front of a camera placed at the water level, with the wings folded and throttle cut manually by a remote operator. While the flight speed cannot be accurately measured, an estimate of the final impact velocity can be made from the final frames of the sequence in which the vehicle flight path is near vertical, with the AquaMAV itself providing a reference length. The vehicle speed was estimated using a software automatic tracking routine (OpenPhysics Tracker) which yielded an approximate impact velocity of 9.1 ms^{-1} (Fig. 7.15B–D). The vehicle was undamaged by the pictured impact, although it is noteworthy that during a later flight an accidental water impact with open wings resulted in the destruction of one of the wing hinges.

Chapter 8
Diving from Flight

Having measured the longitudinal aerodynamics of the AquaMAV in wind tunnel tests (cf. Chap. 7), the data gathered can then be used to analyse the dive performance of the vehicle, as well as estimate and evaluate its dynamic properties. As in Sect. 6.4, we begin by considering a quasi-steady state model, where, furthermore, the aerial and aquatic phases are considered separately and transition phases are omitted. This model is used to obtain planar dive trajectories for both the purely aerial and the purely aquatic phase, providing a clear overview of the achievable performance of the robot. In the second part of the chapter, a more detailed model of the vehicle is developed that accounts for some dynamic effects and explicitly includes the air-to-water transition. The latter model is used to obtain more insight into the vehicle dynamics and into the impact phase. It also serves as a basis for simulations covering several envisaged mission stages.

As in Sect. 6.4, in this chapter the subscripts $_w$, $_f$, $_b$, $_{cg}$ and $_{cb}$ are used to refer to the robot wing, fins, body and centres of gravity and buoyancy, respectively. The coefficient $_t$ is additionally introduced to refer to the tailplane, when the fins are replaced by a conventional tailplane in Sect. 8.2.

8.1 Planar Aerodynamics

The trajectory is defined by velocity and acceleration vectors, $\vec{a}$ and $\vec{v}$, in earth-fixed inertial axes with unit vectors $\hat{X}$, $\hat{Y}$, $\hat{Z}$. Position vectors, $\vec{x}$, are also defined within a robot-fixed reference frame rotated by an angle θ about $\hat{Z}$ from the inertial frame, with its origin at the robot nose and unit vectors $\hat{x}$, $\hat{y}$, $\hat{z}$. The nomenclature for the trajectory model is illustrated in Fig. 8.1.

Coefficients used for the wing, tail and body are taken from the tunnel data shown in Fig. 7.6A–D. The measured force coefficients and centres of pressure are used as look-up tables in the trajectory code, with the data interpolated at each time step. For

R. Zufferey et al., *Between Sea and Sky: Aerial Aquatic Locomotion in Miniature Robots*, Biosystems & Biorobotics 29, https://doi.org/10.1007/978-3-030-89575-4_8

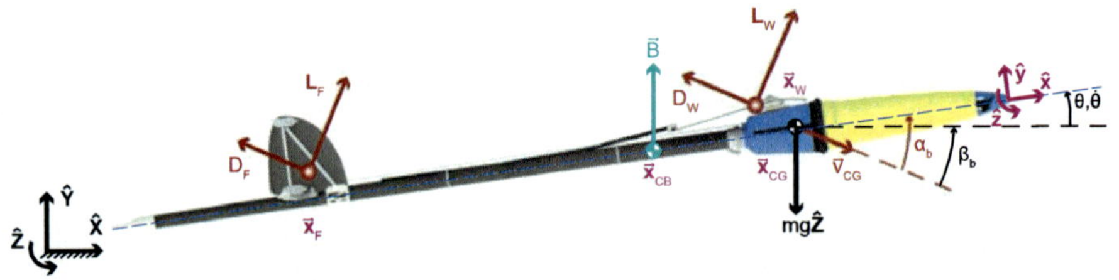

Fig. 8.1 Nomenclature for the equations of motion used in the planar trajectory model, showing the Earth-fixed reference frame ($\hat{X}$, $\hat{Y}$, $\hat{Z}$), the robot fixed reference frame, with its origin at the vehicle nose ($\hat{x}$, $\hat{y}$, $\hat{z}$), and illustrating the definition of body angle of attack, α_{cg}, pitch angle θ and velocity angle with respect to the earth fixed frame, β_{cg}. Also pictured are the wing and tail aerodynamic forces (L_w, D_w and L_f, D_f respectively), the robot weight ($mg\hat{Z}$) and buoyancy ($\vec{B}$)

extrapolation outside of the measured range, the post-stall lift coefficient is used, or linear extrapolation for drag coefficient. However, in these simulations the vehicle was ultimately not simulated reaching an angle of attack outside the measured range, owing to the large tail volume coefficient of the vehicle (36.7%).

The vector forces produced by the wing, $\vec{F}_w$, and the tail fins, $\vec{F}_f$ (Eqs. 8.1 and 8.2), have a lift component perpendicular to the local velocity, $\vec{v}$, and a drag component opposing it, defined by lift and drag coefficients C_{lw} and C_{dw} respectively for the main wing, C_{lf} and C_{df} for the fins, and C_{lb} and C_{db} for the body.

$$\vec{F}_w = \begin{pmatrix} L_w \\ D_w \end{pmatrix} = \frac{1}{2} \begin{pmatrix} C_{lw} \\ C_{dw} \end{pmatrix} \rho_a A_w |\vec{v}_{cg}|^2 \tag{8.1}$$

$$\vec{F}_f = \begin{pmatrix} L_f \\ D_f \end{pmatrix} = \frac{1}{2} \begin{pmatrix} C_{lf} \\ C_{df} \end{pmatrix} \rho_a A_f |\vec{v}_f|^2 \tag{8.2}$$

$$\vec{F}_b = \begin{pmatrix} L_b \\ D_b \end{pmatrix} = \frac{1}{2} \begin{pmatrix} C_{lb} \\ C_{db} \end{pmatrix} \rho_a A_w |\vec{v}_{cg}|^2 \tag{8.3}$$

where A is the component's lifting area, and ρ_a the density of air. The local velocity of the wing and body is taken as the velocity of the centre of gravity, $\vec{v}_{cg}$, but the tail velocity, $\vec{v}_f$ includes the effect of the vehicle's pitch rate, $\dot{\theta}$:

$$\vec{v}_f = \vec{v}_{cg} + \boldsymbol{R}(\pi/2 - \theta)(\dot{\theta}\hat{\boldsymbol{Z}} \times \vec{x}_f) \tag{8.4}$$

where $\boldsymbol{R}$ is a matrix representing rotation about $\hat{\boldsymbol{Z}}$. The angles of attack of the vehicle and its empennage, α_b and α_f respectively, are calculated from the angle the component's velocity in the inertial frame makes with the horizontal, denoted by β:

$$\alpha_{cg} = \theta - \beta_{cg} \tag{8.5}$$
$$\alpha_f = \theta - \beta_f \tag{8.6}$$

β_f includes pitch damping effects (Eq. 8.4), while any pitching damping of the main wing is neglected due to the its proximity to the vehicle centre of mass.

The location of the vehicle's centre of mass, $\vec{x}_{cg}$, and the moment of inertia about that centre, I_{zz}, are computed from accurate CAD models of the vehicle, and are listed in Table 8.1. In the model, wing retraction is again treated as instantaneous, represented by a step change in aerodynamic behaviour and inertia at time $t = t_d$ (in practice the wing takes <0.5 s to retract, depending on airspeed, so any transient effect should have little impact on the overall trajectory).

The forces acting on the robot are resolved into inertial axes, and moments are taken about $\boldsymbol{x}_{cg}$ (Eqs. 8.7 and 8.8):

$$\begin{aligned} m\vec{\boldsymbol{a}} = \boldsymbol{R}(\theta - \alpha_w)\vec{\boldsymbol{F}}_w - m\vec{\boldsymbol{g}} \\ + \boldsymbol{R}(\theta - \alpha_w)\vec{\boldsymbol{F}}_b + \boldsymbol{R}(\theta - \alpha_f)\vec{\boldsymbol{F}}_f \end{aligned} \tag{8.7}$$

$$\begin{aligned} I_{zz}\ddot{\theta}\hat{z} = (\vec{\boldsymbol{x}}_w - \vec{\boldsymbol{x}}_{cg}) \times \boldsymbol{R}(\alpha_w)\vec{\boldsymbol{F}}_w \\ + (\vec{\boldsymbol{x}}_b - \vec{\boldsymbol{x}}_{cg}) \times \boldsymbol{R}(\alpha_w)\vec{\boldsymbol{F}}_b \\ + (\vec{\boldsymbol{x}}_f - \vec{\boldsymbol{x}}_{cg}) \times \boldsymbol{R}(\alpha_f)\vec{\boldsymbol{F}}_f \end{aligned} \tag{8.8}$$

where $\boldsymbol{R}$ is a matrix representing rotation about $\hat{\boldsymbol{Z}}$. These equations are numerically integrated in Matlab using a variable step Runge-Kutta solver.

8.1.1 *Underwater Motion*

While a complete simulation of the vehicle mechanics underwater requires measurement of the forces caused by fluid inertia in rotation and both translation directions, it is nonetheless possible to estimate the ability of the AquaMAV to achieve depth by diving by examining the limiting effects of drag underwater. Neglecting any effect from the increase in Reynolds number, the dimensionless aerodynamic coefficients measured during wind tunnel tests are used here to allow us to simulate motion underwater, replacing the density of air in Eqs. 8.2 and 8.1 with that of fresh water. Due to the presence of buoyancy, a CAD model was used to calculate the total volume displaced by the vehicle, and the centre of displacement ($\vec{x}_{cg}$), which allows us to calculate the torques and moments created by the vehicle's buoyancy,

$$\vec{\boldsymbol{B}} = \rho_w D g \hat{\boldsymbol{Z}} \tag{8.9}$$

Table 8.1 Key parameters for the AquaMAV dynamics computed from a CAD model of the robot. The wings are closed for the majority of the air trajectory simulations and the entirety of the water simulations

Parameter	Value
Distance from nose to centre of gravity (Open Wing)	205 mm
Distance from nose to centre of gravity (Closed Wing)	217 mm
Distance from nose to centre of buoyancy (Closed Wing, empty water tank)	252 mm
I_{zz} (Open Wing)	3.81 kgm^{-2}
I_{zz} (Closed Wing)	4.06 kgm^{-2}
Weight of water displaced, $\rho_w Dg$ (empty water tank)	265 g

where ρ_w is the density of water and D is the volume displaced by the AquaMAV when underwater. The equations of motion then become:

$$\begin{aligned} m\vec{\boldsymbol{a}} =& \vec{\boldsymbol{B}} - mg\hat{\boldsymbol{Z}} + \boldsymbol{R}(\theta - \alpha_w)\vec{\boldsymbol{F}}_w \\ &+ \boldsymbol{R}(\theta - \alpha_w)\vec{\boldsymbol{F}}_b + \boldsymbol{R}(\theta - \alpha_f)\vec{\boldsymbol{F}}_f \end{aligned} \tag{8.10}$$

$$\begin{aligned} I_{zz}\ddot{\theta}\hat{z} =& (\vec{\boldsymbol{x}}_w - \vec{\boldsymbol{x}}_{cg}) \times \boldsymbol{R}(\alpha_w)\vec{\boldsymbol{F}}_w \\ &+ (\vec{\boldsymbol{x}}_b - \vec{\boldsymbol{x}}_{cg}) \times \boldsymbol{R}(\alpha_w)\vec{\boldsymbol{F}}_b \\ &+ (\vec{\boldsymbol{x}}_{cb} - \vec{\boldsymbol{x}}_{cg}) \times \boldsymbol{R}(\theta)\vec{\boldsymbol{B}} \\ &+ (\vec{\boldsymbol{x}}_f - \vec{\boldsymbol{x}}_{cg}) \times \boldsymbol{R}(\alpha_f)\vec{\boldsymbol{F}}_f \end{aligned} \tag{8.11}$$

8.1.2 Dive Trajectories - Aerial Phase

When the AquaMAV wings are retracted completely from a level cruise with wings fully open, the vehicle immediately begins to dive, due to the rearward movement of the centre of gravity and decrease in lift production. Initially, a simulated long dive which allows the vehicle to reach steady state orientation and velocity was examined, initiated at an altitude of 500 m, from level flight at 10 ms^{-1} (Fig. 8.2). 500 m is chosen as a start height as this represents a reasonable maximum operating altitude, given that many countries' civil aviation authorities place a 500 m ceiling on unmanned aircraft operation. Although folding the wings reduces their lift production by 92% (Fig. 7.6C, D), the residual lift has a significant effect on the vehicle trajectory. To gain insight into this effect, this dive is compared with another simulation in which

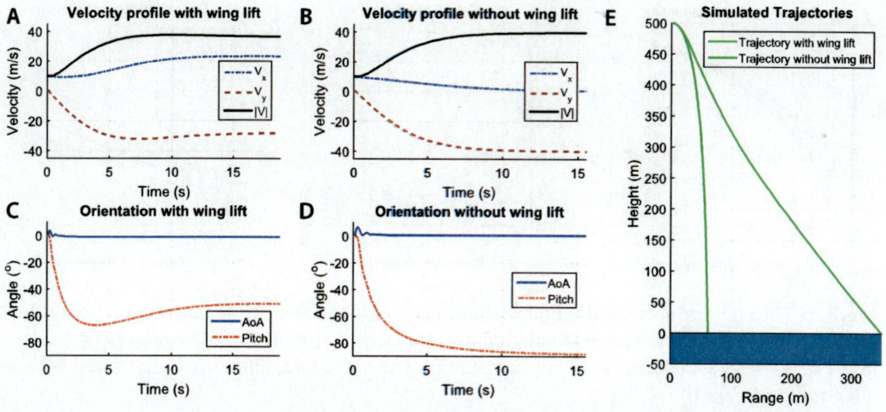

Fig. 8.2 The steady state behaviour of the AquaMAV in an unpowered dive simulated in 3 degrees of freedom using tunnel data, showing the effect of the residual lift of the main wing after folding: (**A**) Velocity profile of the diving AquaMAV. (**B**) Velocity profile of the vehicle if wing lift coefficient is artificially set to zero, but wing drag is retained. (**C**) Pitch and angle attack during a dive. (**D**) Pitch and angle of attack with zero wing lift. (**E**) Dive trajectories in both cases. The plots show that while the wing lift has little effect on the impact velocity, which is largely determined by drag, it has a strong effect on the impact angle

the lift coefficient of the folded wing is artificially set to 0, although the wing drag is retained. In both cases when the wings are folded a short transient pitch oscillation occurs, but is quickly damped out in around 1 s (Fig. 8.2C, D), during which time the vehicle has only descended around 3.5 m. It can be seen from the plots that the residual lift generation of the folded wing has little effect on the vehicle's terminal velocity (36.6 ms^{-1} with lift compared to 39.9 ms^{-1} without, Fig. 8.2A, B), which is largely determined by its drag, but that it does cause the vehicle to reach an equilibrium pitch of $-51°$ from the horizontal (Fig. 8.2C), which means that the vehicle covers significant horizontal distance (Fig. 8.2D). This presents a problem in accurately targeting the impact point of the vehicle in the water, as over long distances wind gusts and turbulence are liable to significantly perturb the trajectory.

Figure 8.2 illustrates the terminal velocity and attitude of the vehicle, but dives are far more likely to be initiated much nearer the surface in practice. Not least because the vehicle reaches an equilibrium angle at terminal velocity in less than 200 m of vertical descent, so diving from any higher has no effect on the final impact other than to increase the distance travelled. Furthermore, at the 36.6 ms^{-1} terminal speed will create significant impact loads, liable to damage the robot.

A fully passive dive, initiated only by the vehicle folding its wings is desirable as it reduces reliance on sensing and control, which is useful when performing an aggressive manoeuvre with a miniature robot. If there is a need to change the water impact velocity or orientation, it would be useful if this could be done by changing the condition at which the wings are retracted, which can be more easily controlled. The effect on final water impact of the speed and altitude when the dive is initiated is therefore examined. Figure 8.3 shows the effect of dive height (the point at which

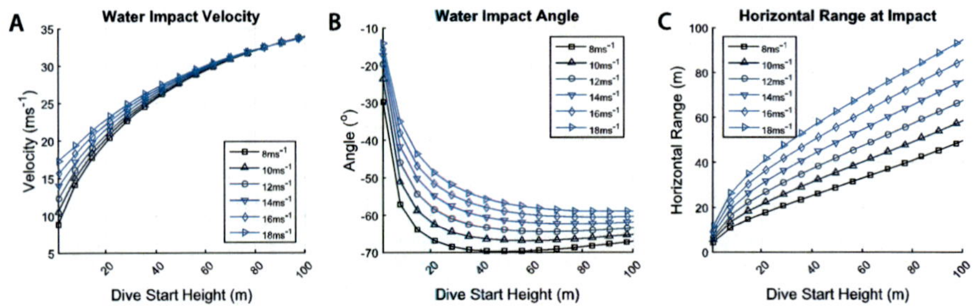

Fig. 8.3 The effect on speed and height when wings are folded on the water impact conditions. Each marker represents one trajectory simulation. (**A**) Impact velocity against dive start height at 8–18 ms^{-1} start speed. (**B**) Angle from horizontal at impact. (**C**) Horizontal distance covered between wing fold and impact

wings are folded) on the final impact of the vehicle, plotted for dives starting at several different cruise speeds. Because the vehicle begins to stall at a lift coefficient of 0.9, its minimum stable speed is approximately 8 ms^{-1}, and this is the lowest speed simulated. The vehicle's maximum propelled flight speed is fundamentally limited by its propulsion. The AquaMAV uses a 130 mm diameter folding propeller, which, when used in combination with the vehicle's 40 W motor, has a an approximate maximum velocity of 18 ms^{-1}, above which the propeller cannot rotate sufficiently fast, and produces drag rather than thrust. This is therefore the maximum start speed simulated.

In Fig. 8.2 it can be seen that the AquaMAV initially pitches more steeply than the equilibrium dive angle, as it must first gather speed. Increasing the initial speed reduces the speed to be gained, and therefore reduces this overshoot. This means that impact angles will be shallower, although horizontal distance covered will be greater. Changing the initial speed has a smaller effect on the impact velocity, which tends asymptotically to terminal velocity as start height is increased. In all cases, the vehicle covers significant horizontal distance before it hits the water, which can be most effectively decreased by reducing the speed at the dive start.

8.1.3 Dive Trajectories - Aquatic Phase

Using the same trajectory model, a simple analysis of the vehicle's motion in water has also been undertaken, by simulating trajectories underwater over a variety of initial conditions. Our aim in this section is to evaluate the extent to which the vehicle drag limits its penetration into water. As such, the significant transient effects the vehicle experiences while impacting the surface, which arise from the rapid accrual of fluid inertia as the vehicle enters the water, are not considered here and the vehicle is treated as starting fully immersed beneath the surface. This is considered reasonable as the neglected inviscid fluid inertia forces are conservative and do not effect total system energy, and other research has been able to reliably predict impact trajectories

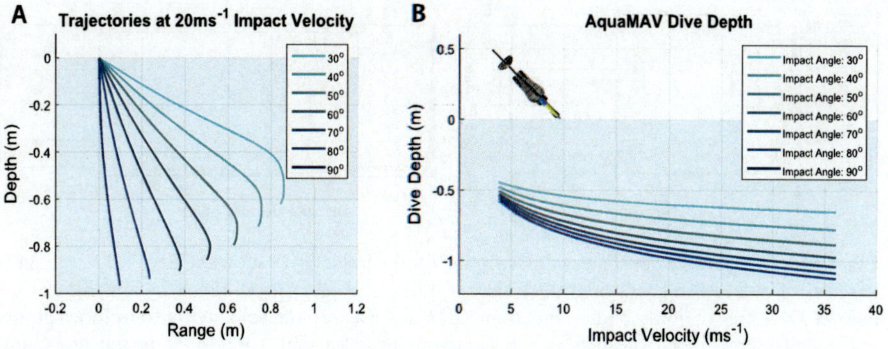

Fig. 8.4 Trajectory of the AquaMAV underwater under changing initial angle and speed: (**A**) Example dive trajectories for impact at 20 ms^{-1}; the AquaMAV does not initially pitch over, but as its speed reduces and the aerodynamic surfaces no longer generate significant force, the rearward centre of buoyancy rotates it nose down. (**B**) Dive depth against impact velocity, plotted for various initial angles. An image of the AquaMAV is included on the same scale as the Y-axis

without considering added mass forces [156], although this relied on artificially fitted coefficients.

The vehicle's tail is hollow which keeps the centre of buoyancy rearward of the wing, and keeps the vehicle stable during a dive. The hollow tail ultimately fills with water through a small check valve, which is then used as propellant to escape the water. The tail filling also shifts the centre of buoyancy forward to make the vehicle float nose up for launch. However, the valve is narrow, and the time taken to fill is much longer than the deceleration time underwater, so the tail is treated as empty throughout. From the dive simulations, it can be seen that the ability of the AquaMAV to achieve significant depth purely through flight momentum is limited. Buoyancy and drag forces arrest the vehicle within approximately 1 m of travel through water. Figure 8.4A shows several example trajectories for impact at 20 m/s at a various angles, with numerical integration stopped when the vehicle's vertical velocity reaches zero. It can be seen that the lifting surfaces keep dive paths fairly straight while the vehicle has some speed, but once the vehicle slows, the nose-down moment from buoyancy in the tail pitches it over. Increasing the impact speed does not yield a significant increase in dive depth, and the vehicle is brought to a stop within 1–2 body lengths below the surface (Fig. 8.4).

8.1.4 Model Assumptions

The trajectory model outlined here assumes that the behaviour of the vehicle is quasi-steady state, and that the wind tunnel measurements can be used to give an accurate description of the vehicle's dynamic behaviour. Because the manoeuvres simulated

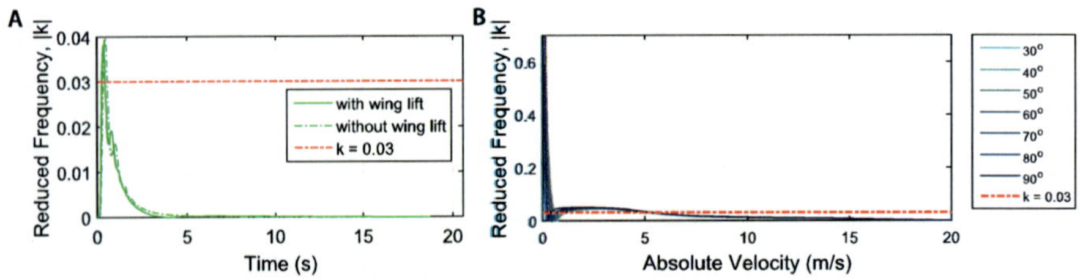

Fig. 8.5 Validity of the quasi-steady assumption for the trajectories shown in Sect. 8.1.2, evaluated in terms of reduced frequency, k. (**A**) k against time for trajectories in Fig. 8.2, showing that k is below 0.04 for the majority of the simulation. (B) k against absolute velocity for trajectories plotted in Fig. 8.4B, showing that the quasi-steady assumption is reasonable for most of the trajectory, until the vehicle's motion is almost arrested, at which point the pitch rate becomes significant relative to the forward velocity

involve rapid changes in velocity and attitude, the validity of this assumption is evaluated by examining the reduced frequency of the vehicle in pitch [157],

$$k = \left| \frac{\dot{\theta} c}{\vec{v}_{cg}} \right| \tag{8.12}$$

where c is the vehicle chord length (0.302 m for the closed wing). Where $|k| < 0.03$, the quasi steady assumption can be considered to hold [154]. Plotting this parameter over the course of the trajectories simulated shows that k remains below 0.04 throughout its trajectory in air, only briefly exceeding 0.03 during the transient pitch oscillation as the wings are folded. In water, k is also low, only becoming large as the vehicle velocity reaches zero at the end of a dive, the portion of the trajectory consisting of slow, buoyancy driven movement, which is of least significance to the vehicle design and operation (Fig. 8.5).

The model also neglects any changes in behaviour due to changes in Reynolds number from the test conditions. The AquaMAV was tested in the wind tunnel at Reynolds numbers between 47,800 and 79,600, and in the water tunnel between 26,000 and 135,000. In the aerial trajectory simulations, the AquaMAV was predicted to reach speeds of up to 36 ms^{-1}, a Reynolds number of 287,000. This would likely result in extensive turbulent flow, which would increase parasitic drag during the dive, as the AquaMAV is highly slender and travels at a small angle of attack during descent, so parasitic drag is likely to be dominated by skin friction. This was borne out by the water tunnel tests, which showed an increase in drag on the fuselage and fins at zero incidence compared to air (Table 7.5). However, further investigation would be needed to quantify this.

8.1.5 Robot Performance

The ability to penetrate the water by diving is strongly a function of scale. As size increases, the significance of fluid resistance (broadly a function of area) in relation to the inertia of a body (broadly a function of volume) declines. The 3 kg Northern Gannet (Fig. 8.6A, B) has been observed reaching depths of up to 10 m in a single dive [158], while the 30 g Common Kingfisher (Fig. 8.6C, D) dives only to around 25 cm [159]. The AquaMAV, weighing 0.2 kg, is predicted to stop under drag within 1 m. The gannet's 15 times larger mass gives it an advantage, and it has more inertia when it hits the water. The gannet also is able to fold itself into a more compact shape, owing to the softness of its feather and tissue. However, there are many effects which have not been explored, including the creation of an air cavity during impact [130], or the use of super-hydrophobic surfaces [139] which could potentially increase the range of future AquaMAV prototypes.

The AquaMAV is also slowed significantly by its hollow tail. When the tail is full, the AquaMAV retains sufficient buoyancy to return to the surface, so the air in the tail is simply extra deceleration. However, without this hollow tail the AquaMAV would be hydrodynamically unstable. It may be possible to make the water tank fill more rapidly, but ultimately the excess buoyancy may be a compromise inherent in the water jet system design.

For animals, diving is a means of prey capture, and the presence of prey determines the necessity of depth. For the AquaMAV, diving into the water serves two purposes. Firstly, it allows the vehicle to penetrate beneath the water in spite of its buoyancy, and without any additional aquatic propulsion. This will allow it to collect water samples beneath the surface, avoiding any floating contaminants, and make other subsurface observations. Secondly, diving directly into the water is an inherently robust landing strategy, relying on passive hydrodynamic stability. This can be compared to 'deep-stall' landings above the surface, or 'soft' landings using floatplanes, both of which require accurate knowledge of the water surface's proximity and the vehicle's attitude. But to gain the advantages of diving directly into the water, folding wings become structural necessity, if the vehicle is to be light enough for efficient flight.

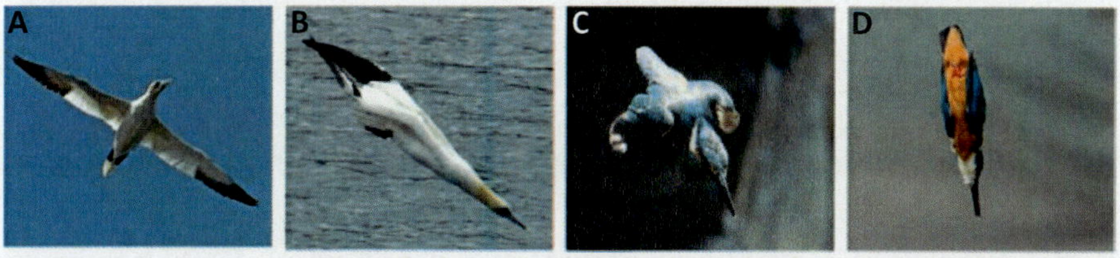

Fig. 8.6 Plunge diving in nature: (**A**) Northern Gannet (*Morus bassanus*, 3 kg mass) in flight, and (**B**) with folded wings about to dive into the water. (**C**) Common Kingfisher (*Alcedo atthis*, 0.03 kg mass) in a dive with wings partially folded and (**D**) with wings fully folded. *Images reproduced under a creative commons license, courtesy of: Andreas Trepte* (**A**), *Mike Pennington* (**B**), *Ryan Cheng* (**C**), *Andy Morffew* (**D**)

The usefulness of diving as a means of water entry is more important in outdoor environments, where the motion of the water surface means that soft landings become more intractable. This is most important for small vehicles, relative to which wave height becomes increasingly large, in an aquatic version of the 'size-grain hypothesis'. Diving directly through the surface is far more insensitive to surface motion than other means of water entry, and it can be seen from Fig. 8.4A that the vehicle initially follows a straight trajectory through the water regardless of the angle at which it impacts.

Passive stability in lieu of active control for landing not only ensures robustness, but also means that robots can be operated more cheaply. However, this comes at the expense of accuracy. From the analysis in Sect. 8.1.2 it is reasonable to suggest that an AquaMAV could autonomously enter a large body of water at a particular point given only a GPS coordinate at which to fold wings, and coarse altimetry. The accuracy with which it hits the water target point will be reduced by the fact that it covers horizontal distance uncontrolled during the dive. If the vehicle is required to dive into more confined spaces, it is important to note that the analysis presented is limited to the longitudinal dynamics of the AquaMAV, and the behaviour of the vehicle where rolling and yawing are permitted has not been investigated. In order to dive vertically down from the start point of the dive without active control, it may be possible to force the vehicle to take a spiral trajectory, which would reduce the horizontal range covered as the vehicle descends. A more controlled and reliable dive manoeuvre would require active control and potentially further actuation possibilities.

In addition to being waterproof, the AquaMAV is adapted specifically for dives into water by its hollow tail, which ensures a stable dive by keeping the centre of buoyancy rearward, but slowly fills with water, orienting the vehicle nose up for take-off after the dive. However, this alone would not ensure penetration beneath the surface, and the ability to fold the wings is essential as it not only protects them from impact but means that the vehicle generates far less lift and drag underwater. A wing large enough to produce lift in flight would produce significantly more force underwater, and force the vehicle off course. The wind tunnel tests of this wing show that the ability to fold does not compromise flight with the wings are open, although the performance of the folding wing is not identical to a rigid equivalent. The folding mechanism has the added advantage of being able to sweep partially backward which expands the vehicle's flight envelope, enhancing the mobility of the prototype.

8.2 System Dynamics Modelling and Simulation

So far in this chapter, we have developed quasi-steady models in order to predict dive trajectories for the AquaMAV and demonstrate the proposed design concept in conjunction with minimal control solutions. The rest of this chapter extends and integrates this work to construct a complete dynamics model for a similar type of vehicle, that includes all operational phases and, especially, detailed consideration of the air-to-water and water-to-air transition phases including time-varying effects. We also introduce further actuation possibilities and explicitly consider the dynamics of the system.

As evident from Chap. 4, only few of the existing aerial-aquatic robot studies demonstrate the full mission spectrum on a prototype [109, 113, 160], and the complete realisation of fully-functional, controlled and automated vehicles capable of efficiently performing full missions remains challenging. An important step towards this goal is to improve the robots' control and navigation capabilities, allowing for automated, safe and eventually autonomous operation. To support this process, a comprehensive understanding of the dynamics of such MAVs is required. Dynamic modelling studies are still relatively scarce and typically focused on rotary-wing robots [110]. Particularly more unconventional designs have not yet been the subject of extensive modelling work.

In the following sections we develop a simple multi-phase model of the dynamics of a folding-wing plunge-diving vehicle, with the aim of supporting the development of navigation and control strategies, facilitating the investigation of minor design changes, and providing further insight into the dynamics of bioinspired aquatic MAVs. The simulated vehicle is based on the AquaMAV and exploits a combination of variable wing-folding and conventional actuation and propulsion mechanisms to operate effectively in water and air, as well as transition effectively between the two domains. The dynamics model presented comprises three basic components, representing, respectively, the aerial, underwater and transition phases of operation. The resulting model provides a detailed representation of the MAV, while also retaining a relatively simple structure, amenable for practical applications. The model is used to simulate and analyse the dynamics of the (type of) vehicle studied in all phases of operation. By providing both insight and a simulation framework, the developed model constitutes a useful basis for advanced control and navigation work on the unconventional type of vehicle studied.

8.2.1 Simulated Vehicle Properties and Preliminary Considerations

The aim here is to conduct a somewhat generic analysis applicable to different forms of the same type of plunge-diving vehicle, therefore the details of the platform are no longer of primary importance. To obtain quantitative results, however, the AquaMAV geometry used in the first part of this chapter and introduced in Chap. 7 continues to be used as a baseline, and the aerodynamic model is still based on the previously discussed wind tunnel data (cf. Chap. 7). Nonetheless, to allow for fuller consideration of the system dynamics, a number of changes are introduced to the simulated robot. These include the following: (1) inclusion of conventional control surfaces on the tailplane, (2) conventional T-tail instead of V-tail, (3) absence of the water thrust mechanism for water exit, (4) consideration of an epicyclic gearing mechanism [8] (Sect. 6.1), allowing for effective propulsion in both air and water.

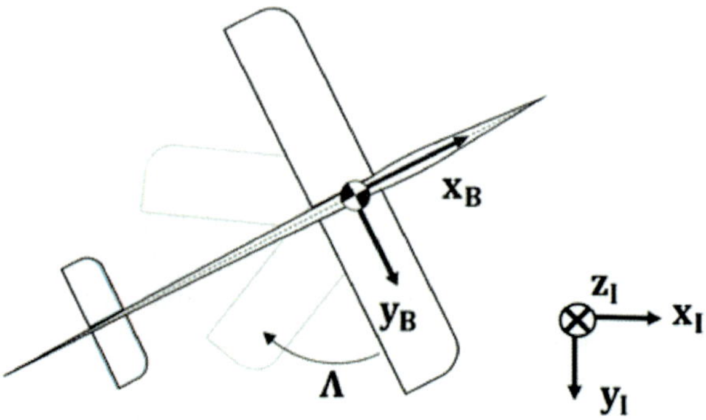

Fig. 8.7 Schematic of the vehicle used as baseline for simulation, upon which many of the modelling assumptions were based. The top view schematic shows the body-fixed coordinate frame, inertial coordinate frame and definition of the sweep angle Λ [6]

(1) was deemed necessary to allow for future control and navigation work, the main intended application of the devised model. (2) was introduced to facilitate the actuation by decoupling the longitudinal and lateral control. (3) was introduced because the ejection approach is still the subject of further studies and hence excluding the water thruster makes the simulation both simpler and more general. With applications in mind, it is considered simpler to use conventional propellers to extract the vehicle from the water. A simple schematic of the simulated vehicle is shown in Fig. 8.7.

The simulated vehicle is assumed to have completely foldable elliptic wings, a wing span of approximately 0.6 m and a mass of approximately 0.2 kg. It can be controlled by means of the wing folding (0–90°), through conventional elevator and rudder surfaces in the tailplane (assumed span: 0.19 m, root chord: 0.05 m), and through thrust variations. The subsequent analysis remains focused on the longitudinal dynamics.

To facilitate flight dynamics analysis, the model developed in the subsequent section is expressed in a body-fixed coordinate system, centred at the instantaneous CG of the body and aligned with the body axis, as clarified by Fig. 8.7. Note that when the wings fold, several changes occur in the vehicle configuration/geometry, i.e. the wing area, wing chord and wing span change. Furthermore, the CG shifts backwards, and hence the inertia moments change. Simple linear functions were used to approximate these changes.

The body is assumed to be symmetrical around the longitudinal (x_B) axis, and the wings move symmetrically and only in the xy plane. Hence the combined CG of the two wings only moves in x-direction, as the changes in y direction cancel out between the left and right wing. Using the new wing CG, the total CG of the body can be computed, and from it the moments of inertia.

Note that while several assumptions were made in the modelling process, the main goal of this study was not to develop a highly accurate model of a particular platform but to construct a somewhat flexible model of the type of MAV design. The obtained simulation framework will allow for control and navigation approaches to be designed and tested, and, when used in practice, the model can be adapted to the actual design studied.

8.3 Modelling

The overall model is comprised by several components, which are discussed further in the rest of this section. To narrow down the problem, the focus is placed on the longitudinal dynamics and forces. Many of the key functionalities and features can be analysed to a large extent in a vertical plane, before adding the complexity of the lateral component. Furthermore, the lateral aerodynamic forces were not characterised in the wind tunnel. Given the extensive range of sweep angles considered, existing studies do not provide a complete and reliable overview of the behaviour of the vehicle throughout this range.

The equations of motion (EOM) for the longitudinal dynamics can be defined in a conventional way, according to standard aeronautics notation, with additional terms to cover the impact and underwater phases. The equations below are a combined reformulation of Eqs. 8.7, 8.8, 8.10 and 8.11 in a body coordinate frame, with the additional inclusion of added mass forces, thrust force and cross-coupling terms. The equations cover all operation phases (air, water, transition), with some terms taking zero values in some phases.[1] For better readability, we introduce some new notation for axis-aligned components of the variables, as clarified below. The notation chosen is aligned with standard aeronautics notation. Moreover, note that capitalised subscripts are used to denote types of forces, as opposed to the vehicle parts they refer to, which continue to be referenced using lower-case letters.

$$\dot{q} = \frac{1}{I_{yy}}\left[M - F_B(x_{bary} - x_{CG})\cos\theta + T z_{CG}\right] \tag{8.13}$$

$$\dot{u} = \frac{X}{m} - qw - g\sin\theta + \frac{F_B\sin\theta}{m} - \frac{F_{AM,u}}{m} + \frac{T}{m} \tag{8.14}$$

$$\dot{w} = \frac{Z}{m} + qu + g\cos\phi\cos\theta - \frac{\cos\theta F_B}{m} - \frac{F_{AM,w}}{m} \tag{8.15}$$

$$\dot{\theta} = q \tag{8.16}$$

Here u and w are the translational velocities along the x_B and z_B axes; q is the body pitch rate, Θ is the pitch attitude, m is the body mass; I_{yy} is the body moment of inertia about the y axis; X and Z are the aerodynamic forces in x_B and z_B direction; M is the aerodynamic moment about the y_B axis; F_B is the buoyancy force; g is the acceleration due to gravity; T is the thrust force; F_{AM} are the added mass force terms; T is the thrust force from the propeller; and x_{CG} and y_{CG} describe the position of the CG along the respective body-frame axes, with respect to the vehicle nose.

The forces in the above equations are discussed in the remainder of this section, with emphasis on the unconventional components arising from the unconventional design of the MAV studied, such as 0° to 90° folding wings and air-water transitioning.

[1] This implies aerial-aquatic robots can be considered hybrid dynamic systems.

8.3.1 Aerodynamic Forces and Flight Phase

In view of the envisaged application of the aerodynamic model for simulation and control work, a smooth global model was developed to represent the vehicle aerodynamics. A system identification process was used for this, based on wind tunnel force measurements obtained on a similar type of platform [6] (cf. Sect. 8.2.1). Using experimental data to represent the aerodynamics, as opposed to analytical approximations, allowed us to model the effects of the extreme wing sweep changes in a realistic way, as well as to consider possible side effects related to the mechanical design, e.g. the wing-folding mechanism, which involves parts of the wings sliding beneath each other [6]. While such effects are platform specific, and may vary significantly even between different realisations of the same platform, they are representative of the typical issues that may arise in comparable vehicles. As such, it can be useful, at least on a qualitative plane, to model these effects.

In the aforementioned wind tunnel tests, the aerodynamic coefficients of the separate components of the MAV (body, tail, wings) were measured at different angles of attack (AOA) ($-2°$ to $14°$ in intervals of $2°$) and, where relevant, different wing sweep angles ($0°$ to $90°$ in intervals of $10°$) [6]. The aerodynamic coefficients for the longitudinal forces and moments were thus assumed to depend on the AOA and wing sweep angle only. Reducing the aerodynamics to the most dominant effects was considered sufficient for a first approximation, although considering additional effects, e.g. angular rate effects, may prove necessary if rapid manoeuvres are considered. Thanks to separate measurements performed on the different vehicle components, each component could be modelled separately, which is helpful when modelling the water-air transition phases (cf. Sect. 8.3.3), where different parts of the body enter/exit the water at different instants.

Except for the large-range wing-folding, the MAV during flight is a conventional fixed-wing MAV. The body and tail coefficients were thus found to be accurately approximated by simple second-order functions of the AOA, with coefficients obtained via linear regression. Note that, as mentioned in Sect. 8.2.1, the wind tunnel data was obtained for a V-tail whereas the simulated vehicle was assumed to have a conventional T-tail: it was assumed, as a first approximation, that the simulated tail would generate the same overall aerodynamic forces as the V-tail characterised in the wind tunnel.

By contrast, the wing coefficients were found to vary in a more complex way, due to the large range of sweep angles considered. While several aircraft designs include sweeping wings, the variation in sweep is typically confined to a smaller range. It is also typical for these vehicles to only operate at two or three discrete sweep configurations, rather than exploiting the full range. In this sense, the folding mechanism in the AquaMAV is unconventional, both due to the covered range and due to the possible utilisation of this range. While the final design may eventually involve switching only between a limited number of sweep angles, a full use of the available sweep range may also be considered. Hence, the model was required to provide full and continuous coverage of the sweep range.

The wing aerodynamic coefficients were found to depend nonlinearly on the AOA and the sweep angle. The nonlinearity in the wind tunnel data may also have been accentuated by the folding mechanism. While it is difficult to distinguish between aerodynamic and mechanical effects/disturbances, it was considered of interest to capture the most evident irregularities in the model, as discussed above.

To obtain a flexible and smooth global model without resorting to high-order model structures, we used a model identification approach based on multivariate simplex B-splines. This approach allows for accurate models of highly complex datasets to be obtained by combining low-order local models. The underlying idea is to divide the identification domain into a net of non-overlapping simplices (triangulation), identify a local model for each simplex using the datapoints contained within it, and enforce continuity conditions of the desired order between neighbouring simplices. In particular, each simplex t_j in the triangulation has a local barycentric coordinate system and supports a local polynomial of the form:

$$p(x) = \begin{cases} \sum_{|\kappa|=d} c_\kappa^{t_j} B_\kappa^d(b_{t_j}(x)) \text{ for } x \in t_j, \\ 0 \text{ for } x \notin t_j \end{cases} \tag{8.17}$$

where $c_\kappa^{t_j}$ are the coefficients of the polynomial (B-coefficients), b_{t_j} are the barycentric coordinates of point x with respect to simplex t_j, d is the degree of the polynomial, $\kappa = (\kappa_0, \kappa_1, \ldots, \kappa_n)$ is a multi-index containing all permutations that sum up to d, and $B_\kappa^d(b)$ are the local basis functions of the multivariate spline:

$$B_\kappa^d(b) := \frac{d!}{\kappa!} b^\kappa, \quad d \geq 0, \kappa \in \mathbb{N}^{n+1}, b \geq 0 \tag{8.18}$$

In a system identification context, the model structure thus locally takes the form of Eq. 8.17, with the B-coefficients representing the model parameters and the basis functions being the independent variables used to model the system output. Standard estimators can then be used to estimate the parameters. An in-depth discussion of this identification approach can be found in Ref. [161].

The outlined approach was applied, with an ordinary least squares estimator, to obtain a model for the wing lift and drag coefficients, respectively, as a function of the wing sweep and AOA. The model structure and triangulation used for each model were based on an evaluation of the wind tunnel data, and fine-tuned in order to obtain an accurate and plausible result using the lowest possible model degree and smallest number of coefficients. Given that the model was intended to be used for control and generally within optimisation schemes, at least first order continuity was considered necessary. The chosen model structures and their accuracy are presented in Table 8.2, while Figs. 8.8 and 8.9 show the obtained models and original datapoints—it can be seen that a high accuracy is achieved and a high proportion of the data are captured, providing a realistic basis for the current simulation.

In summary, the aerodynamic coefficients of the MAV take the following form:

Table 8.2 Modelling parameters and resulting model performance for wing aerodynamic coefficients

Output	Model definition			Model performance		
	degree	continuity	# simplices	RMS	RMS [%]	R^2
C_L	3	2	8	0.06	2.4	0.99
C_D	3	2	8	0.02	3.8	0.99

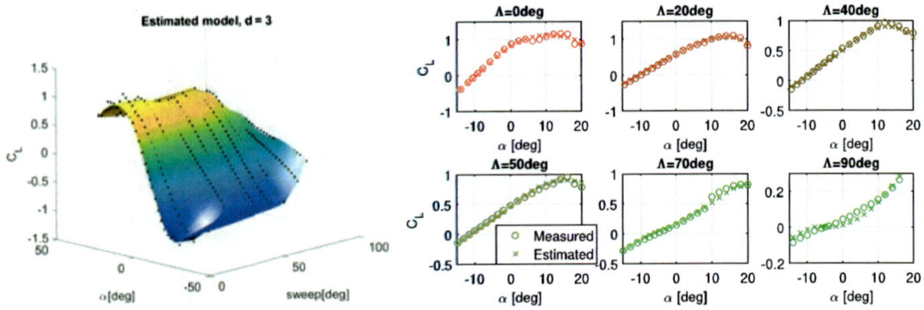

(a) Measured wind tunnel data (dots) and model (surface)
(b) Wind tunnel data versus model, for different sweep angles Λ

Fig. 8.8 3rd order spline model for $C_{L,wing}$

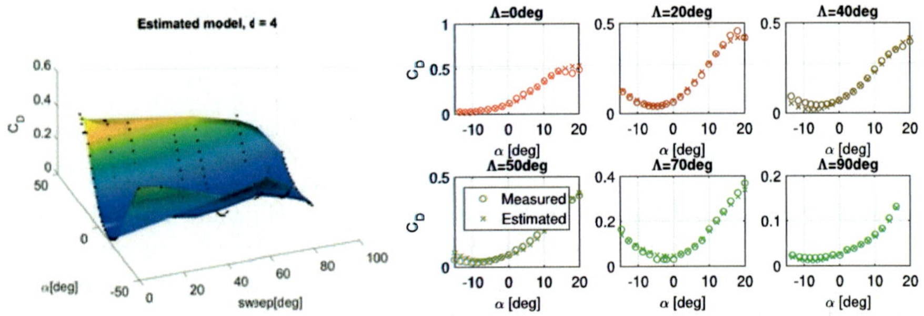

(a) Measured wind tunnel data (dots) and model (surface)
(b) Wind tunnel data versus model, for different sweep angles Λ

Fig. 8.9 3rd order spline model for $C_{D,wing}$

$$C_{lw} = f^3(\alpha, \Lambda), \qquad C_{dw} = f^3(\alpha, \Lambda) \tag{8.19}$$

$$C_{lt} = f^2(\alpha_t), \qquad C_{dt} = f^2(\alpha_t) \tag{8.20}$$

$$C_{lb} = f^2(\alpha), \qquad C_{db} = f^2(\alpha) \tag{8.21}$$

where the subscripts w, t and b denote, respectively, the wings, the tail and the body. The lift and drag forces for each component i were then computed according to the standard quasi-steady aerodynamics formulation, cf. Eq. 8.1. As previously, note that when computing the tail forces, the velocity and AOA in the lift and drag equations consider the effect of the body pitch rates. Expressing these variables in terms of the body velocities gives:

$$V_t := |\vec{v}_t| = \sqrt{u_t^2 + w_t^2} \tag{8.22}$$

$$\alpha_t = \arctan\left(\frac{w_t}{u_t}\right) \tag{8.23}$$

$$\text{with} \quad w_t = w + q(x_t - x_{cg}), \quad u_t = u_b = u_{cg}$$

where u and w are the velocities in x_B and z_B direction, respectively, x_t is the tail position along the x_B axis.

The moments due to the wings and tail are then given by:

$$M_t = (x_{cp,t} - x_{cg})Z_t \tag{8.24}$$

$$M_w = (x_{cp,w} - x_{cg})Z_w \tag{8.25}$$

where the force application points, indicated by the subscript cp (centre of pressure), were likewise modelled from the wind tunnel data as simple functions of the AOA (and sweep angle). The tail force in the above expression includes the effect of potential elevator deflections.

In addition to the basic lifting surfaces, the vehicle was assumed to have conventional control surfaces, as well as a propeller. The elevator effectiveness was estimated based on the assumed vehicle and tail geometry [162], and refined qualitatively considering the output to standard input signals—as for example shown in Fig. 8.11b, which presents the response to a small elevator doublet. The propeller was approximated in a simplified way, based on the maximum static thrust predicted by the analysis in Ref. [8].

8.3.2 Underwater Phase

In the underwater phase, the vehicle dynamics are determined by hydrodynamic, static and added mass forces, as well as propulsive and actuating forces introduced by the propeller and control surfaces. Based on the envisaged vehicle design, the wings were assumed to remain folded at all times while underwater.

The hydrodynamic forces were considered to be analogous to the aerodynamic forces, with the air density ($\rho_a = 1.225\,\text{kg/m}^3$) in Eqs. 8.1–8.3 being replaced by water density ($\rho_w = 997\,\text{kg/m}^3$).

The buoyancy forces depend on the MAV volume. To facilitate consideration of the transition phases, for this model the body geometry was approximated as a simple symmetric solid of revolution, with the cross section approximating the vehicle shape. In particular, as can be seen in Fig. 8.7, the body is wider at the front, where the payload is located, while the hind part is essentially a thin rod. The total buoyancy was approximately equal to the CAD-determined value used in the previous trajectory modelling.

Lastly, underwater operation may involve non-negligible added mass forces. In aerial operation, added mass effects are negligible owing to the very low density of air. Even underwater, added mass terms are often considered comparatively negligible—however, due to the high fluid density, this is not always the case, depending on the specific geometry and mode of operation of a vehicle. While studies on UAVs and small submarines often neglect the added mass component altogether, other studies have suggested that this component can be significant, particularly for bodies transitioning between different fluids [163]. Given that the added mass is in any case non-negligible during water-air transitions, the term was also retained for the underwater component of the model.

Added mass forces are challenging to measure in isolation, as they typically occur in conjunction with other forces. Deriving accurate analytical expressions for these forces can likewise be challenging, particularly for objects with relatively complex geometries, which are typically characterised using either CFD or experimental methods [164, 165]. The alternative is simplification, i.e. using elementary geometries to represent a body. After impact, the AquaMAV resembles a slender ellipsoid combined with a thin plate. Even though the wings may contribute less to the added mass than the rest of the body, the interaction between the different body parts may be significant. However, this effect is also challenging to capture. It has been suggested that the added mass of bodies consisting of combined elementary shapes can be obtained by summing the added masses of each sub-component, but that that this approximation sometimes leads to significant errorsspsciteBrennen1982 and hence is not necessarily more reliable than other, simpler approaches.

In view of the above considerations, the equivalent ellipsoid approach was adopted to represent the added mass forces. As discussed in Ref. [167], the added mass of a slender body (length > width) can be roughly approximated by the added mass of an ellipsoid with the same volume and the same length. The added mass forces are conventionally defined as,

$$F_{AM,x} = \lambda_{11}\dot{u} \tag{8.26}$$

$$F_{AM,z} = \lambda_{33}\dot{w} \tag{8.27}$$

$$M_{AM,y} = \lambda_{55}\dot{q} \tag{8.28}$$

where λ_{ij} are the added mass coefficients, representing the added mass in direction i due to an acceleration in direction j. In view of the assumed symmetry of the vehicle, cross-coupling terms (where $i \neq j$) are considered negligible. In an intuitive sense, the added mass can be viewed as the mass of fluid displaced in each direction. The added mass coefficients of the equivalent ellipsoid along each relevant axis can then be obtained according to following approximations [167]:

$$\lambda_{11} := m_v k_{11}, \quad \text{where } k_{11} = \frac{A}{2-A} \tag{8.29}$$

$$\lambda_{33} := m_v k_{33}, \quad \text{where } k_{33} = \frac{C}{2-C} \tag{8.30}$$

$$\lambda_{55} := I_v k_{55}, \quad \text{where } k_{55} = \frac{(a^2-c^2)^2(A-C)}{(2(c^4-a^4)+(C-A)(c^2+a^2)^2)} \tag{8.31}$$

where a and b are, respectively, the long and short semi-axes of the ellipsoid longitudinal cross-section, A and C are parameters depending on the geometry of the ellipsoid (see Ref. [167] for details), and m_v and I_v are the total virtual mass and inertia. As the MAV body was assumed to be approximately rotationally symmetric, the minor semi-axes of the ellipsoid were assumed to be equal and both denoted by b. m_v and I_v are thus:

$$m_v = \frac{4}{3}\rho\pi ab^2 \tag{8.32}$$

$$I_v = \frac{4}{15}\rho\pi ab^2(a^2+b^2) \tag{8.33}$$

8.3.3 *Air-Water Transitions*

The air-water and water-air transitions constitute the most challenging phase to model, representing highly complex multi-phase flow problems. During water entry, the violent impact at high velocities further adds to the complexity: as the body impacts the considerably more dense water, the flow is heavily disrupted and huge forces act on the body. As the two problems can be formulated in a similar way, we discuss the water entry case here and only point out the main differences considered in the water exit case.

The impact phase sees considerably large forces acting in a time-varying way over an extremely short period of time. The moment the vehicle comes into contact with the water surface, the drag forces increase drastically and buoyancy and added mass forces are introduced into the system. Furthermore, the forces are influenced by the continual variation in submerged body mass, surface area and volume of the body, as it transitions from one phase to the other, and by the resulting variation in impact geometry. Moreover, while the MAV is approximately axissymmetrical, it does not have an ideal shape, as even in a folded arrangement the wings and tail do not fully conform to the streamlined shape of the rest of the vehicle.

In the literature, the fluid-dynamic problem of water entry has been widely studied since the initial formulation by Wagner [168]. Even now, many studies are based on assumptions, e.g. vertical impact, ideal symmetric shape, ideal fluid at rest, etc. [163, 169]. While several more elaborate analytical formulations have been introduced [170], e.g. accounting for additional effects such as cavitation, or for complex geometries or kinematics, especially applied studies typically use CFD approaches or significant simplifications to simulate water entry [163]. Given that the intended purpose of the current simulation model was to support the development of control systems and to provide a basic idea of the system dynamics, computationally heavy models, e.g. CFD-based ones, were not considered. Instead, the water entry was modelled in a highly simplified way. Nonetheless, an additional aim of the current study was to evaluate to what extent a highly simplified model is acceptable and how realistic the predicted outcomes are.

Thus, in addition to the assumptions introduced in Sect. 8.3.2 for the full submersion case, a number of additional assumptions were made specifically regarding the water entry process (and exit process, where relevant). In particular, while the impact was not constrained to the ideal vertical case, the impact surface was assumed to be analogous to the surface that would be obtained during a vertical impact—as clarified in Fig. 8.10. This allowed for the same simplified geometry to be assumed for the body throughout the impact phase, with the impacting surface approximated by a circular cross section (with varying width). Moreover, the instantaneous volume and surface in air/water could be more easily computed. Besides implicitly in the state and force components, the impact angle was considered in terms of the submerged body length, which was defined to be aligned with the body axis. All states were computed considering the impact angle.

The buoyancy force was computed as for the underwater case, with the instantaneous submerged volume replacing the volume of the full body. Similarly, the approximate barycentre position was computed considering the submerged part of the body.

The added mass forces were based on the same simplifications discussed in Sect. 8.3.2, extended to the time-varying case. Comparable semi-empirical approaches have for instance been applied to missiles exiting water [163], and to conic wedges entering water [169]—while the current study involves more extensive simplifications, this suggests that the type of approach is in principle applicable to water entry/exit problems. Unlike in the underwater case, the geometry of the body varies during the transition phases. Assuming the added mass in air is negligible, the total added mass can be assumed to depend only on the submerged part of the body. Hence, the added mass coefficients λ_{ij} are no longer constant, but depend on the instantaneous geometry of the current equivalent ellipsoid. Furthermore, the submerged volume of the body changes during water entry/exit, influencing the virtual mass and inertia

terms. Hence, the total added mass (Eqs. 8.26–8.28) was reformulated to include an additional term [163, 169]:

$$F_{am,x,impact} = \lambda_{11}\dot{u} + \dot{\lambda}_{11}u \tag{8.34}$$

$$F_{am,z,impact} = \lambda_{33}\dot{w} + \dot{\lambda}_{33}w \tag{8.35}$$

$$M_{am,y,impact} = \lambda_{55}\dot{q} + \dot{\lambda}_{55}q \tag{8.36}$$

Recalling the previous definition of the added mass coefficients given in Eqs. 8.29–8.31, the derivatives $\dot{\lambda}_{ij}$ can be expressed as,

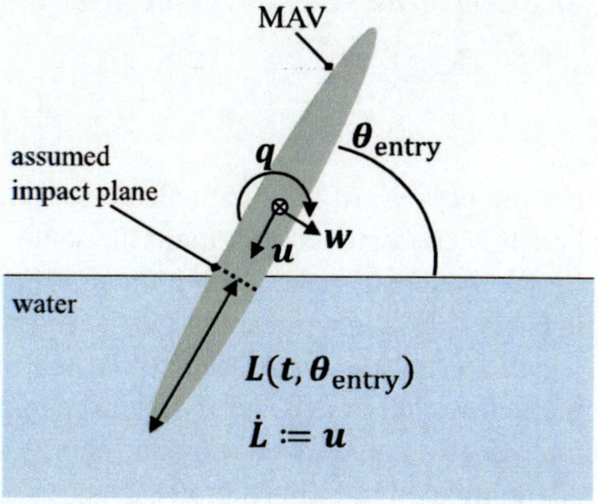

Fig. 8.10 Impact geometry clarification. Note: the terminology is analogous for the water exit case, with the vehicle being reversed (such that velocity u points upwards)

$$\dot{\lambda}_{11} = \dot{k}_{11} m_f + \dot{m}_f k_{11} \tag{8.37}$$

$$\dot{\lambda}_{33} = \dot{k}_{33} m_f + \dot{m}_f k_{33} \tag{8.38}$$

$$\dot{\lambda}_{55} = \dot{k}_{55} I_f + \dot{I}_f k_{55} \tag{8.39}$$

where the virtual mass and inertia derivatives are given by

$$\dot{m}_v = \frac{4}{3}\dot{L}\rho\pi b^2 \tag{8.40}$$

$$\dot{I}_v = \frac{4}{15}\pi\rho b^2 \dot{L}(3a^2 + b^2) \tag{8.41}$$

where $\dot{L}$ is the rate of change of the submerged length. By definition, in the current problem $\dot{L} = \dot{u}$, as the length is aligned with the body x-axis. The coefficient derivatives were obtained by analytical differentiation of Eqs. 8.29–8.31 (which, through the coefficients A and C depend on the instantaneous length submerged).

For the water entry case, the above formulation neglects potential cavitation or air entrainment effects. The influence of upsurge was also neglected based on preliminary tests which suggested the overall influence on the impact forces was relatively small. Note that the above approximations break down in the first instants after impact (and the final instants at water exit), as here the submerged geometry no longer resembles a slender body, but more closely represents a spheroid. To deal with this phase, the ratio a/b was assumed to remain constant during these infinitesimal times—implying the submerged part of the body is approximated as a cone.

Finally, the drag (and potentially lift) forces were modelled. Two different approaches were considered. Firstly, it is possible, as a rough approximation, to assume—as in Sect. 8.3.2 for the completely submerged case—that the hydrodynamic forces take the same form as the aerodynamic forces in air, with only the density changing. Taking a similar approach as in Ref. [171], the submerged part of each component of the vehicle was determined based on the submersion geometry, and the force produced by each component was scaled by the submerged area. Thus, for example, the wing forces are given by,

$$F_{w,transition} = \frac{1}{2}V^2 C_{lw}(\rho_w S_w(t) + \rho_a S_a(t)) \tag{8.42}$$

For the purpose of moment calculations, the force application points for each component were assumed to remain the same. The above expression may be acceptable for the water exit case. During the impact phase, however, the flow around the body is likely to be highly chaotic. Furthermore, cavitation or air entrainment may occur, which for instance may lead to the tail, and potentially parts of the wings, being ineffective during and shortly after the impact. In this sense, it was considered more realistic to assume that during and shortly after the impact, the hydrodynamic forces are dominated by skin friction effects, as also suggested in Ref. [171]. Hence, in the alternative formulation we assume—as in Ref. [171]—that the drag force coefficient is given by:

$$C_d = \frac{0.0307}{Re^{1/7}} \tag{8.43}$$

where the Reynolds number is computed considering the submerged length of the body.

The above formulations provide an estimate of the forces occurring at water entry and exit, nonetheless they are highly simplified and may not be sufficiently realistic in practice. This will be discussed further when evaluating the results.

8.4 Simulation Results and System Analysis

The developed model allowed us to simulate and analyse the dynamics of the envisaged aquatic MAV in different phases of operation (cf. Fig. 7.1), deriving basic insight and requirements for control and effective overall operation. Particular emphasis was placed on the most relevant and unconventional manoeuvres, i.e. plunge-diving and air-water-air transitions. This section presents examples of simulation results, illustrating the full mission cycle and allowing for the plausibility of the model to be evaluated, followed by a brief appraisal of the system dynamics, with an outlook on control work. Finally, the transition phases are briefly considered in further detail.

8.4.1 Simulation of Different Phases

Figures 8.11, 8.12 and 8.13 show a simulation of the different phases of operation of the modelled MAV, covering (i) steady and manoeuvring flight, (ii) steep dive manoeuvre, (iii) water impact and entry, (iv) steady underwater operation, (v) return

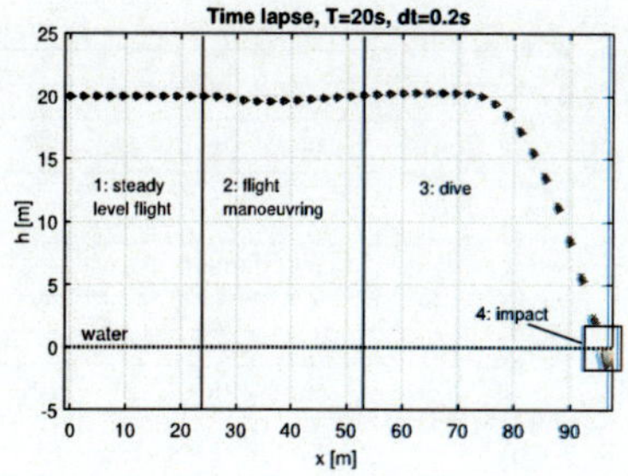

(a) Time lapse showing trajectory in longitudinal plane, body not to scale

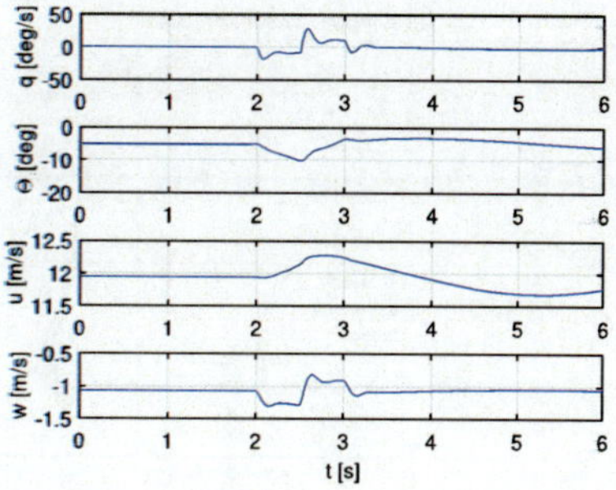

(b) State evolution, part 2: doublet manoeuvre in air]

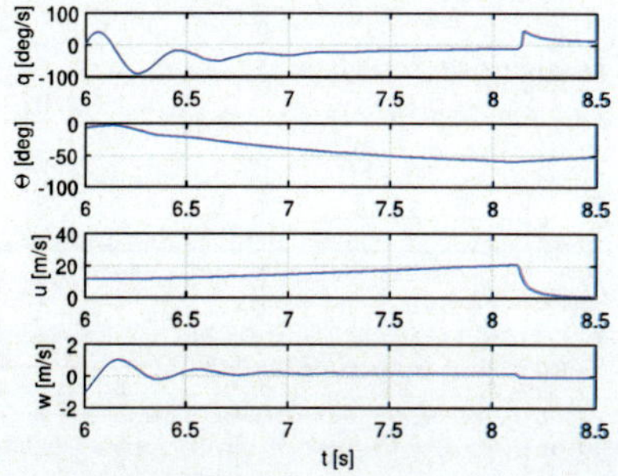

(c) State evolution, part 3: wing folding-initiated dive and water impact

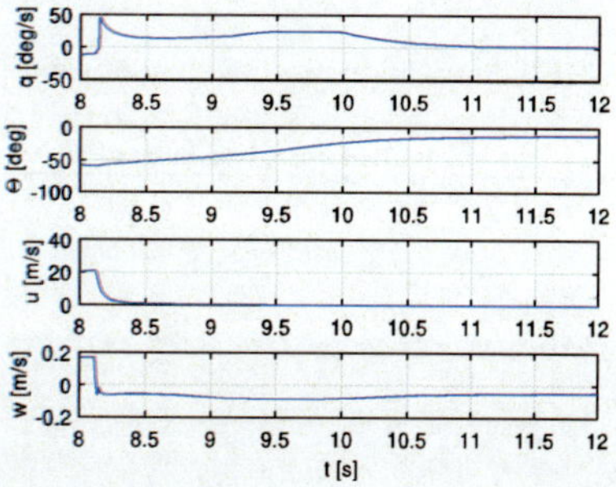

(d) State evolution, part 4: detail of water impact and transition

Fig. 8.11 Simulation example: aerial phase and water impact

to surface and water exit. The obtained results suggest that the devised model can effectively represent both the purely aerial aerial and aquatic phases, and the transitions from air to water and vice versa. The resulting system behaviour is physically plausible and agrees well with experimental results obtained previously on similar vehicles [6]. It must be noted that, while all phases are covered, in order to achieve a continuous, full mission cycle, active stabilisation is needed for some parts of the cycle, e.g. stable water exit and transition back to flight, and transition from water impact to controlled underwater operation. The control component will be the subject of future work and is not discussed further here.

8.4.2 System Dynamics: Aerial and Aquatic Operation

To obtain more in-depth insight into the behaviour of the MAV, its operational envelope was determined and its dynamics were evaluated across the aforementioned envelope. Figure 8.14 shows examples of numerically $\alpha - V$ trim curves obtained for the vehicle at a number of different wing sweep angles, in air and in water. In the interest of completeness, the underwater curve for open (unfolded) wings is also

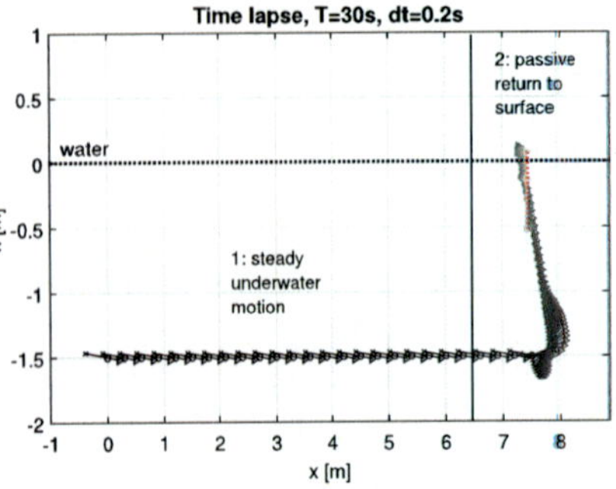

(a) Time lapse showing trajectory in longitudinal (xz) plane, body not to scale

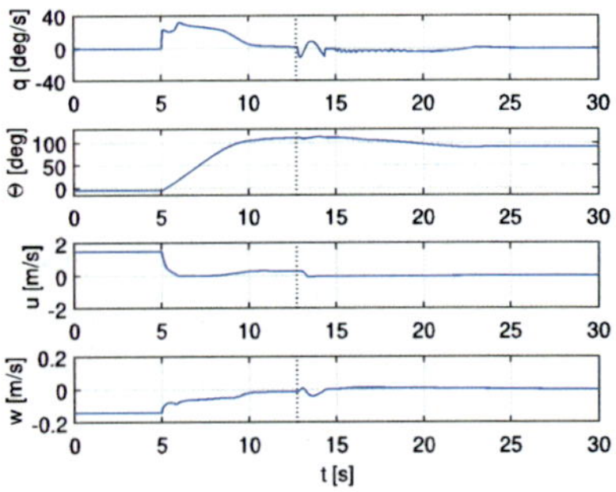

(b) State evolution

Fig. 8.12 Simulation example: underwater phase and passive rise to surface

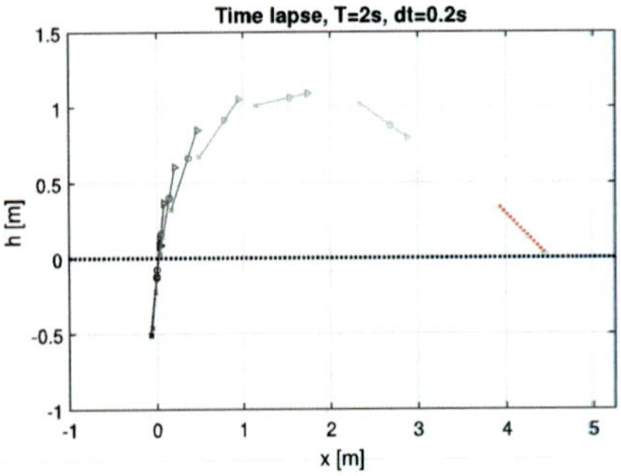

(a) Time lapse showing trajectory in longitudinal (xz) plane, body not to scale

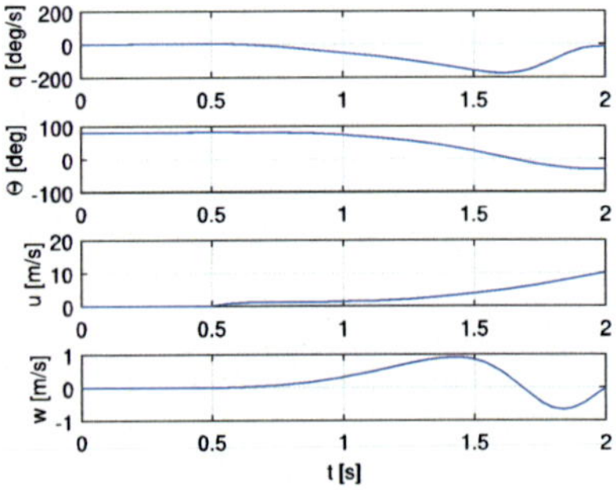

(b) State evolution

Fig. 8.13 Simulation example: unstabilised launch from floating condition

shown, despite the envisaged vehicle operating with closed wings while underwater. The obtained curves are qualitatively plausible, with a lower angle of attack being required to stabilise at higher speeds (note that the extensive negative angle of attack range is a result of the wing aerodynamics, cf. Sect. 11.5). The typical operating speed in air and water was found to be, respectively, ~10 m/s and ~1 m/s. The combination of wing sweep and elevator input leads to a fairly large envelope especially in flight. It is also interesting to note that the behaviour of the system is nonlinear with the sweep angle: changes are initially very gradual, then increasingly rapid, with the flight envelope changing rapidly between $\Lambda = 60°$ and $\Lambda = 90°$.

In the underwater phase the operational envelope is in a sense more restricted, as the wings remain folded. The envisaged epicyclic gearbox [8] allows for adequate thrust production underwater: with the current setup, it was found to be possible to maintain the vehicle at a constant depth at velocities of 1.0–2.5 m/s. Holding a constant horizontal position, on the other hand, is virtually unachievable in practice due to the relatively limited underwater actuation possibilities. A neutrally buoyant

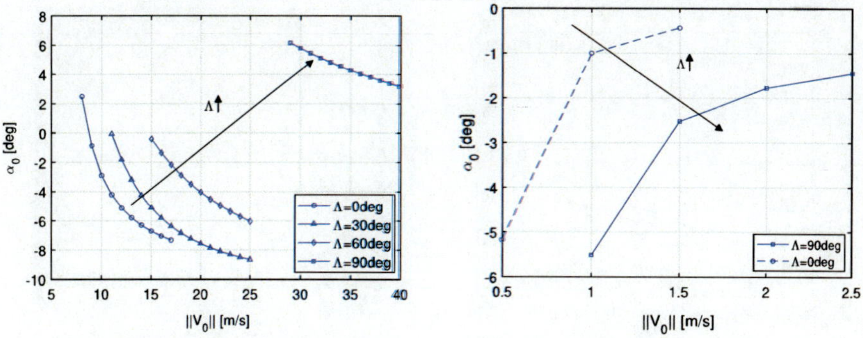

(a) Steady level flight trim curve

(b) Steady underwater operation trim curve.

Fig. 8.14 Steady operational envelope in air and water, considering different wing sweeps

alternative configuration, or some form of water surface vehicle would be better suited for missions requiring extended periods of time at an approximately constant position. Alternatively, more elaborate or additional underwater actuation mechanisms, e.g. a form of buoyancy control, may allow for more effective underwater manoeuvring and/or horizontal position-keeping. These options will be considered in future work.

To obtain further insight into the system dynamics, stability was evaluated at the trim points defining the stable operational envelope. This approach can be considered an acceptable first approximation for the steady aerial and underwater phases, though it cannot account for aggressive nonlinear manoeuvres. The nonlinear model (Eqs. 8.13–8.16) was thus linearised around a set of steady conditions, based on the previously discussed trim results, and the resulting local linear models, of the form $\dot{\mathbf{x}} = \mathbf{A}\mathbf{x} + \mathbf{B}\mathbf{u}$ (with $\mathbf{x} := [q, \theta, u, w]^T$, $\mathbf{u} = [\delta_E, \delta_T, \Lambda]^T$), were used for stability analysis of the aerial and aquatic phases. Stability during the transition phases, where the system becomes inherently time-varying, was not considered at this stage.

Figure 8.15 shows the eigenvalues computed for a representative number of linearised models, representing different aerial and aquatic equilibrium conditions. As expected, the system dynamics are considerably influenced by the significant changes in configuration (wing-folding) and, especially, mode of operation (air/water). Clearly, one of the most significant challenges of the studied system is thus the *combination* of very different components, even if the single components may, on their own, be relatively uncomplicated. This also suggests that some form of control adaptation or switching may be required to allow for a fully-functional and autonomous vehicle.

In the aerial phase (Fig. 8.15a, b), the system dynamics are conventional for the open-wings case, with two complex pole pairs that approximately correspond to short period and phugoid modes. With increasing sweep angle, the stability increases, with both pole pairs moving to the left in the complex plane. The slower mode additionally increases somewhat in damping, moving slowly towards the real axis with little change in frequency, while the faster mode increases significantly in frequency, with a small decrease in damping only at very high sweep angles. At low sweep angles,

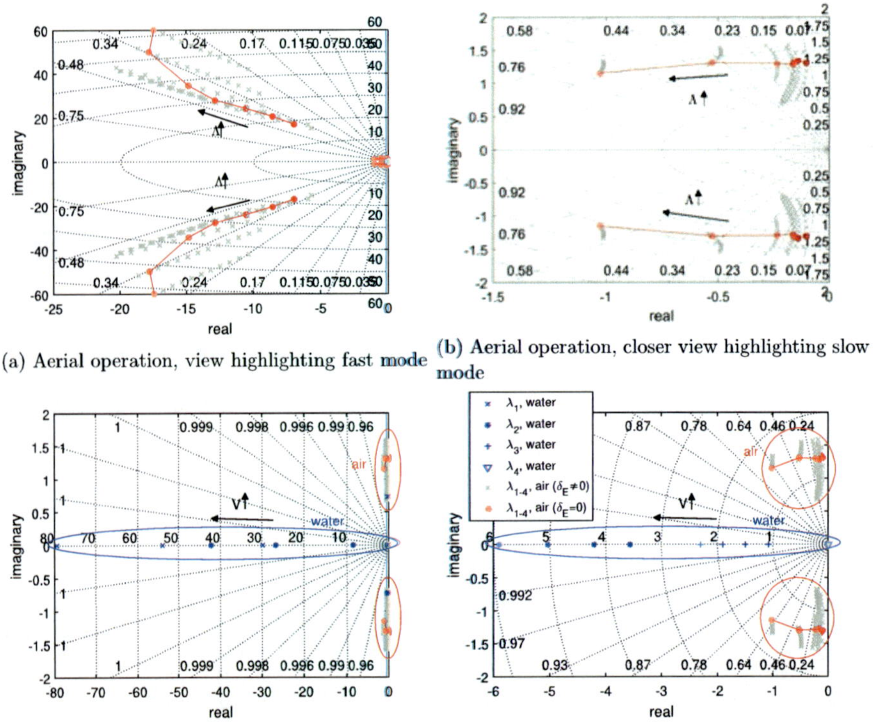

(a) Aerial operation, view highlighting fast mode

(b) Aerial operation, closer view highlighting slow mode

(c) View highlighting fast modes of underwater operation

(d) View highlighting slower modes of underwater operation

Fig. 8.15 Eigenvalues λ_i of the system in aerial and underwater operation, in the vicinity of different steady operating conditions. For the aerial component, the red dots show the neutral-elevator case, while the grey crosses show the larger flight envelope achievable with additional controls. The underwater component is shown only for the case of closed wings and includes both throttle and elevator input

the slow mode is quite poorly damped. It is also interesting to observe that, echoing the trim curves, changes are gradual until sweep angles of approximately 60° and then more rapid, presumably as a combined effect of the aerodynamics and the folding mechanism, which affects the wing profile. The underwater phase is shown in Fig. 8.15c, d for the relevant folded-wings case. All eigenvalues were found to be aperiodic and stable. While the system is therefore stable, it must be borne in mind that the MAV is slightly positively buoyant and hence requires continual control inputs to remain underwater.

In addition to steady operation at constant altitude (or depth), an additional phase that is interesting to consider is the unpowered dive. Technically part of the aerial operation, this manoeuvre is nonetheless unconventional and constitutes a central element in the envisaged MAV's mission cycle as it enables the robot to enter the water. Figure 8.16 shows the steady glide conditions achievable with different wing sweep angles. As already suggested in Ref. [6] for a predecessor robot, the

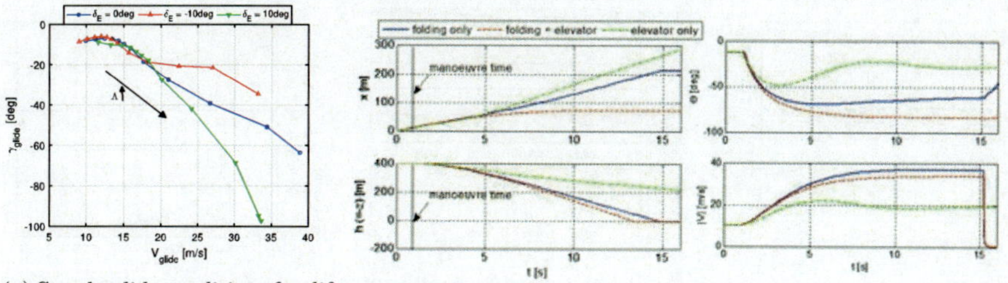

(a) Steady glide conditions for different sweep angles and elevator trim inputs

(b) Example of large-scale dive manoeuvre using only wing-folding, only elevator input, or both.

Fig. 8.16 Unpowered steady glide conditions and dive manoeuvre example, showing the glide angle (≈pitch attitude) and velocity magnitude attainable by changing the wing sweep only, or both the wing sweep and the elevator deflection angle

wing-folding considerable affects the resulting trajectory and can potentially be used as sole mechanism to carry out a dive. Folding back the wings leads to terminal impact angles of up to ~70°—as the wings still generate lift when folded, the steady-state pitch attitude cannot be vertical. However, the additional use of conventional control surfaces potentially allows for larger impact angles to be attained, at slightly lower velocities. Given that steeper entry angles are generally preferred, as they allow for greater depth to be reached, reduce the potential damage of the impact, and lead to more predictable underwater trajectories, using conventional actuators in addition to the folding wings may be desirable. Nonetheless, the dive manoeuvre requires a degree of compromise. Achieving steep entry angles requires high velocities, which may be damaging to the MAV and furthermore lead to it covering large distances before reaching the desired attitude. During this time the robot is also less easily controlled due to the high speed and smaller lifting surfaces. In this sense, it is not advisable to reach terminal speeds, but rather to use a combination of actuator inputs to attain an acceptable non-steady entry condition.

8.4.3 *Outlook: Analysis of Transition Phases*

The analysis in the previous section focused on the two main phases of operation of the studied MAV, however the main challenge is given by the far shorter transitions *between* these phases. Thanks to the full inclusion of time-varying effects, the developed model can be used to study the transition phases in some detail, also allowing for a full study of the stability, especially during the slower water exit manoeuvres. A brief overview of initial simulation results is given in this section as an example.

Figures 8.17 and 8.18 show the changes in the forces and states occurring during and immediately after water impact, for different entry angles and velocities. It can be observed that while the forces on the vehicle are huge, i.e. predicted to reach several hundred g, the time scale of the process is very short compared to the remainder

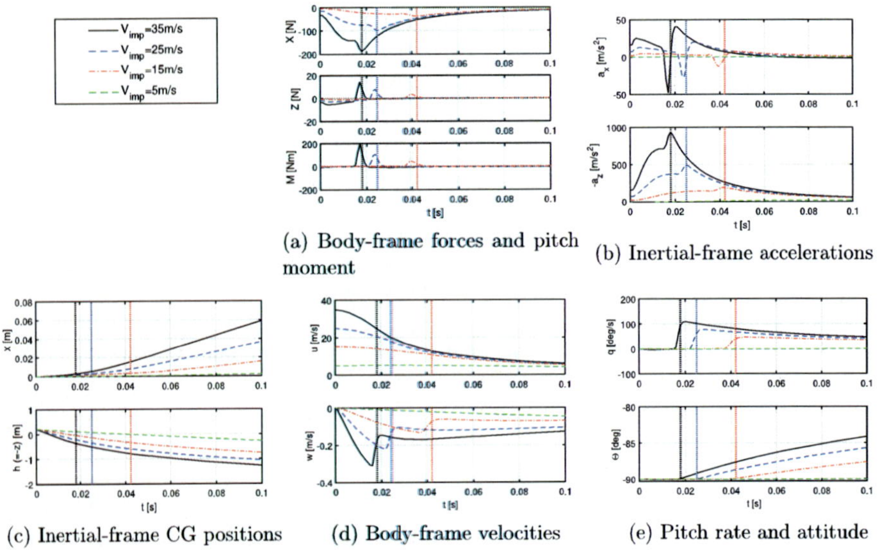

(a) Body-frame forces and pitch moment
(b) Inertial-frame accelerations
(c) Inertial-frame CG positions
(d) Body-frame velocities
(e) Pitch rate and attitude

Fig. 8.17 Water impact process for different body-aligned entry velocities V_{imp} (vertical entry angle); the dashed lines indicate the time at which the body is fully submerged

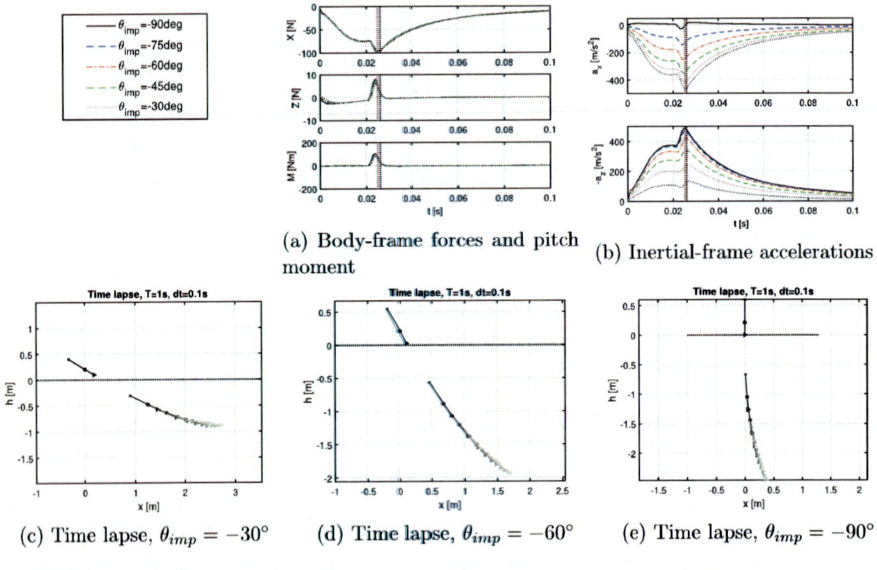

(a) Body-frame forces and pitch moment
(b) Inertial-frame accelerations
(c) Time lapse, $\theta_{imp} = -30°$
(d) Time lapse, $\theta_{imp} = -60°$
(e) Time lapse, $\theta_{imp} = -90°$

Fig. 8.18 Water impact process for different entry angles θ_{imp} (25 m/s entry velocity); the dashed lines indicate the time at which the body is fully submerged; the bottom plots show time lapses over 1 s of the system dynamics. This suggests that small changes in the forces (e.g. due to slightly different impact model formulations) may have a relatively limited overall effect, especially on the resulting attitude and position. Moreover, given that the test conditions in any realistic scenario will be unideal and very difficult to predict or characterise, it is likely that the real system will deviate non-negligibly from the modelled system. In particular, the uncertainty deriving from external disturbances, as well as manufacturing inaccuracies, flexibility effects, etc., may be larger than the uncertainty in the model. Another consequence of the small time scale and very large changes occurring in the system is that even the current highly simplified representation of the transition phases requires a very small integration step for meaningful simulation results to be obtained, which adversely affects simulation speed and may limit online applicability. The combination of the above points suggests that aiming for an extremely high accuracy may not be necessary—instead, it is advisable to develop a controller that can handle a degree of model uncertainty and disturbance occurring during and immediately after the impact phase. For simulation purposes, a possible approach may be to represent the impact using a surrogate term acting only at a single instant, but approximating the time-varying forces of the entire impact process.

In terms of impact conditions, unsurprisingly, steeper impact angles and higher impact velocities allow for greater depths to be reached, with shallow angles and low velocities eventually leading to failure of the body to fully penetrate the water surface. The most noticeable effect of the initial impact velocity is on the depth and on the normal velocity component w (Fig. 8.17), while the attitude obviously most influences the trajectory occurring after impact (Fig. 8.18). The predicted depths (cf. Fig. 8.18) attained are slightly higher than previous estimates that neglected unsteady effects [6], but remain in a comparable range.

The water exit phase, driven by a powerful propeller, is somewhat slower and less aggressive than the impact phase. Here the time-varying evolution of the forces is in a sense more important because it influences the feasible exit trajectories. While the exit phase will need to be further studied, with particular emphasis on the stability of the *time-varying* system, our model provides intial insight, allowing for the time-varying evolution of the forces and states to be estimated. An example of this is shown in Fig. 8.19, which presents the forces and positions obtained for a range of different exit angles, under the assumption that such angles can be achieved through underwater manoeuvring. As expected (and illustrated by Fig. 8.19c–e), a more vertical exit angle tends to facilitate the exit process. Nonetheless, considering that the vehicle needs to transition back to horizontal flight once out of the water, it may be advantageous to exit the water at an angle. In this sense, it may be interesting, in future work, to consider different exit manoeuvre options, i.e. exiting from a near-vertical floating condition after passively rising to the surface, or commencing the exit manoeuvre underwater, hereby potentially exploiting some underwater velocity as well as buoyancy. An important point to consider in this evaluation is the effect of wetness on the efficiency of the propeller and the lifting surfaces, which at present was neglected.

Like the impact process, albeit to a lesser extent, the exit process is relatively fast and difficult to model accurately. Being less rapid and violent, it is however slightly

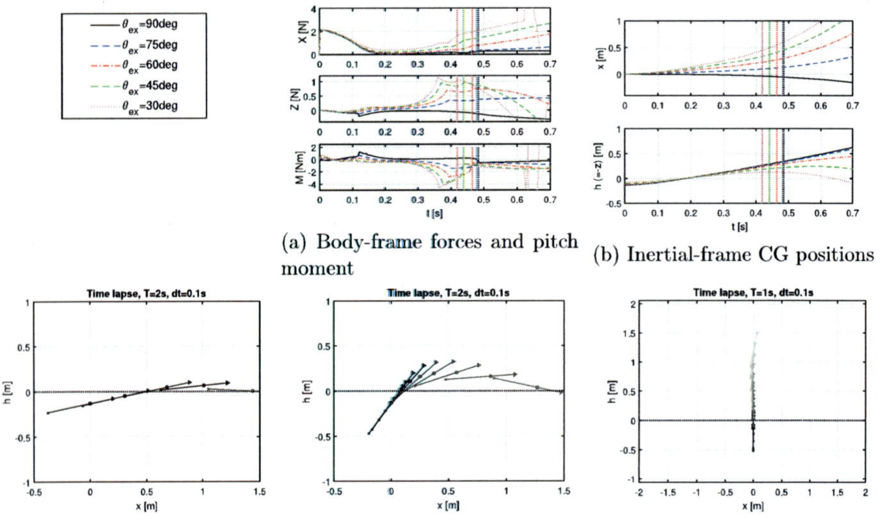

(a) Body-frame forces and pitch moment

(b) Inertial-frame CG positions

(c) $\theta_{ex} = 15°$

(d) $\theta_{ex} = 60°$

(e) $\theta_{ex} = 90°$

Fig. 8.19 Water exit process for different exit angles θ_{ex}, starting from near-floating conditions; the dashed lines indicate the time at which the body is fully out of the water (which for some conditions is never the case); the bottom plots show time lapses over 2 s

more sensitive to inaccuracies. Furthermore, given that the exit process requires stabilisation, an accurate representation of the system is of greater relevance.

Nonetheless, despite significant differences in time scales and force magnitudes, the transitions in both directions share several characteristics. Considering that a high accuracy is challenging to achieve and the current model involves a number of simplifications, it is thus important, firstly, to attempt an experimental validation of the obtained results, and, secondly, to implement a robust control strategy for the exit process that can handle a degree of model uncertainty.

8.5 Concluding Remarks on Simulation

In this chapter, two models were presented for a bird-inspired aerial-aquatic robot, the AquaMAV. The first focused on characterising typical dive trajectories of the robot and yields a valuable overview of the achievable performance of folding wings-based designs. Trajectory simulations elucidate the terminal velocity and orientation of the vehicle during and dive, and show the range over which the robot can affect its final impact conditions by changing its cruise speed or altitude. A passively initiated dive is shown to be suitable for entering confined bodies of water, and the vehicle's underwater stability means that structural stresses from hydrodynamic loads will be greatly reduced.

Building on the initial model, a more detailed second model is developed to fully characterise the robot dynamics and provide a simulation framework. The second model includes aerial, underwater and transition phases of operation, and hence allows for all parts of the envisaged vehicle behaviour to be simulated. The aerodynamic forces were modelled using a system identification approach based on wind tunnel data, which led to an accurate low-order model fully accounting for the morphing wing design. The remaining forces were characterised with simplified analytical models. Particular emphasis was placed on the air/water transitions, which were represented under full consideration of time-varying effects.

The resulting model was found to yield a detailed and plausible representation of the MAV, and retains a relatively simple structure, amenable for practical applications. The model was used to analyse the behaviour of the vehicle in different operation phases. The system dynamics were shown to vary considerably, due to both the wing-folding and the air-water transitioning, however the vehicle was found to be stable within the flight envelope considered and outside of the transition phases. Basic characterisation of the transition phases illustrated the forces occurring during water entry and exit and highlighted the difficulties of simulating these phases accurately, owing to the very large forces and short time scales. While a highly accurate representation may be unnecessary due to the short time scales, it is considered advisable to conduct an in-depth stability analysis of the time-varying components, and to consider modelling uncertainty when developing controllers for the transitions. This is particularly relevant for the more gradual water exit phase.

The final model represents a useful basis for further simulation and control work, e.g. providing an initial testing possibility for an unconventional type of vehicle, and allowing for model-based control methods to be applied.

Chapter 9
Aquatic Escape: Repeatable Escape with Combustion

9.1 Water Reactive Chemistry

Several systems have been developed with aerial-aquatic locomotion capabilities but without demonstrating consecutive transitions to flight from water. Moreover, while some multirotor vehicles possess the ability to operate in both air and water [108, 109], the transition to flight is typically constrained to very calm sea conditions. Fixed-wing robots able to transition dynamically between water and air through high-power thrust bursts represent a low-cost, versatile and more reliable solution. Compared to multirotor vehicles, this approach that would simultaneously result in an increased flight range and allow for aquatic escape in a wider variety of conditions.

The transition from water to air is a complex task. Water is three orders of magnitude denser than air, with profound effects on the drag and buoyancy forces in play. As such, lifting surfaces such as propellers and wings cannot operate efficiently in both media, which makes it difficult to accelerate a partially submerged vehicle to flight velocity. Take-off from water typically requires the aircraft to position its lift and propulsion surfaces well clear of the water to develop aerodynamic forces. A small vehicle, however, is constantly threatened by the possibility of water surface motion immersing its aerodynamic surfaces, with waves typically being large relative to its size [117]. A method for producing a high amount of power for a short period of time that is insensitive to water immersion would solve this issue, allowing a vehicle to quickly escape the surface and transition to flight at a safer altitude (Fig. 9.1).

In nature, transition to flight is shown by several animals [172, 173]. For example, many fish are also able to shortly leave the surface through accelerating underwater to speeds of 10 m/s and leveraging ground effect for glides above the water surface [68]. This behavior has however not yet been demonstrated by robotic platforms with fin propulsion due to power limitations [174]. Birds also demonstrate capability of water exit after plunge-diving manoeuvres [46]. The flying squid achieves full take-off by rapidly compressing its membrane, producing high power jetting which is followed by extension of its fins to glide back to the surface [77]. This jetting occurs by radial contraction of its cavity, driven not only by muscle tissue, but also

R. Zufferey et al., *Between Sea and Sky: Aerial Aquatic Locomotion in Miniature Robots*, Biosystems & Biorobotics 29, https://doi.org/10.1007/978-3-030-89575-4_9

Fig. 9.1 Concept view of the Aquatic Jump-glider escaping the water

by energy stored in stretched collagen fibres. This allows the squid to produce the high power needed for impulsive aquatic escape.

Impulsive aquatic escape manoeuvres necessitate large power reserves for rapid, aggressive acceleration. In robotics, this is often addressed with the use of elastic energy storage, compressed gas, or combustible fuels, all of which can rapidly release energy for actuation. A number of studies have proposed thrusters using compressed air [1] or liquid-nitrogen [175], however these were limited to a single use and did not include gliding capability. Other relevant research has leveraged and studied jet propulsion for transitionless, underwater locomotion [176, 177].

9.1.1 Combustion

Combustion offers a means of creating high pressures without the need for a pressurised tank and release mechanism. Churaman et al. for instance created robots weighing only 314 mg which were able to jump over 8 cm vertically using explosive actuators [178]. Similarly, researchers have recently shown take-off of the 100 mg Robobee by means of hydrogen explosions [179]. While both of these systems showed high-power thrust, they required an off-board power supply to function and did not benefit from gliding to prolong the jump. Larger, untethered robots that use combustion to power soft terrestrial jumping have also been built [180, 181]. These systems use pressurised liquid reservoirs of butane gas as fuel, metered into an elastic silicone combustion chamber by electronic valves, and ignited by a spark. However, the provision of multiple liquid fuel tanks and flow control apparatus result in a significant increase in mass and complexity of the thruster, which would be impractical to integrate in an small-scale aerial-aquatic vehicle.

9.1.2 Solid Reactants as a Combustion Gas Source

Large volumes of fuel gas can be stored compactly as liquid under pressure. However, this requires storage and regulation systems able to sustain the large pressures required, which adds complexity and makes components heavier. The provision of a pressurised container of combustible gas is also a hazard in and of itself. If the combustible fuel is instead produced by the reaction of two separately stable components, high pressures can be avoided, and the fuel storage and dispensing systems can be greatly simplified. The use of solid compounds for gas storage is common in many applications. One example is the use of sodium azide (NaN_3) decomposition to release nitrogen (N_2) for car airbag deployment. More recently solid alkali metal hydrides have been explored by the fuel cell industry as a compact means of hydrogen storage.

More relevant to aquatic propulsion are the studies undertaken by the US Navy [182] examining Lithium Hydrides for use in a torpedo propulsion system, or the work by [183] examining solid reactants for buoyancy control. Moreover, the usage of water as a reactant is an attractive possibility for aerial-aquatic robots as their environment can be exploited in order to reduce system mass.

The fuelling and ignition system phases in which droplets of water are drawn from the surrounding water body and injected into a small container of calcium carbide powder is shown in Fig. 9.2. This method permits multiple, separate, reactions through the distribution of water droplets. As such, the total fuel required for the mission can be safely stored in the single container and recharged in a matter of seconds. Counting the water droplets provides a good estimation of the amount of acetylene produced. Acetylene and calcium hydroxide are produced through the following exothermic reaction:

$$\mathrm{CaC_2 + H_2O \rightarrow C_2H_2 + Ca(OH)_2} \tag{9.1}$$

The acetylene gas (C_2H_2) is allowed to escape into the partially flooded combustion chamber where it mixes with air. For this process each milligram of (CaC_2) yields 0.3 mg of (C_2H_2), for a (CaC_2) purity of 75%. Lastly, the gas mix is ignited by an electric arc and combustion occurs:

$$\mathrm{2C_2H_2 + 5O_2 \rightarrow 4CO_2 + 2H_2O} \tag{9.2}$$

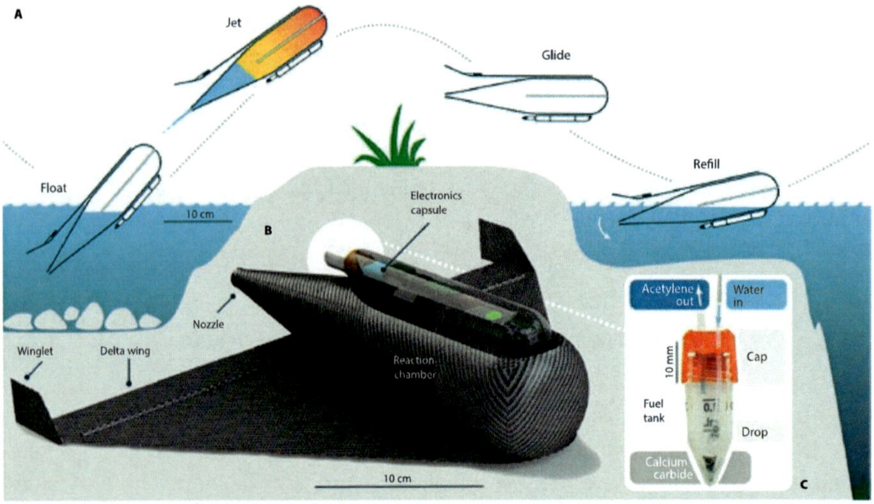

Fig. 9.2 Mission profile, robot operation and design. (**A**) The proposed mission stages, showcasing the transition from a floating state to an airborne jetting phase and back to floating. (**B**) 3D model render of the underside of the robot highlighting key features. (**C**) Section of fuel container, with fuel for one mission cycle, showing a water drop about to react with the calcium carbide

This reaction is highly exothermic, and the temperature and pressure increase rapidly in the chamber (Fig. 9.3). This forces the water through a nozzle, out of the combustion chamber, producing the required thrust.

9.2 Simulation and Validation with Fixed Experiment

9.2.1 Physics of Jet-Gliding

It is necessary to develop an analytic physics model to gain insight on the pressure evolution and flight trajectory, and to determine various robot parameters and their impact on the robot's performance (Fig. 9.4).

There are two leading physical aspects to the pressure-propelled vehicle problem. Firstly, an external view of the robot explains the flight trajectory based on the forces on the robot. Secondly, internal considerations permit calculation of the pressure evolution and thrust. Both sets of equation have to be solved simultaneously as the system is accelerating.

External considerations The equations of motion applied to the system yield Eq. 9.3.

$$\Sigma \vec{F} = \frac{d\vec{P}}{dt} = \vec{T} - \vec{W} - \vec{D} - \vec{L} \tag{9.3}$$

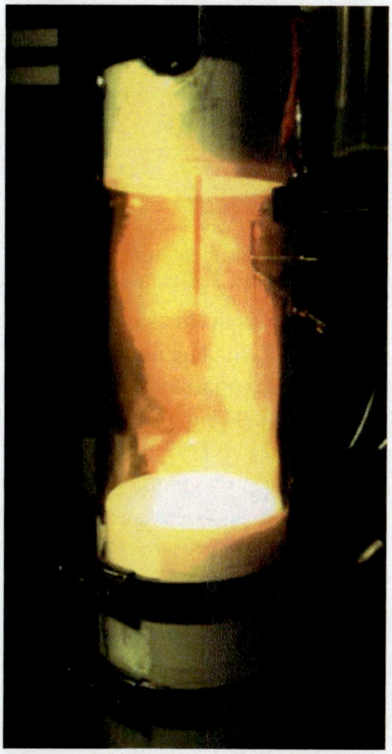

Fig. 9.3 Photo of the acetylene combustion process occurring in a transparent chamber. The electrodes are visible on the right side. The chamber is attached by the two lower buckles to a force sensor

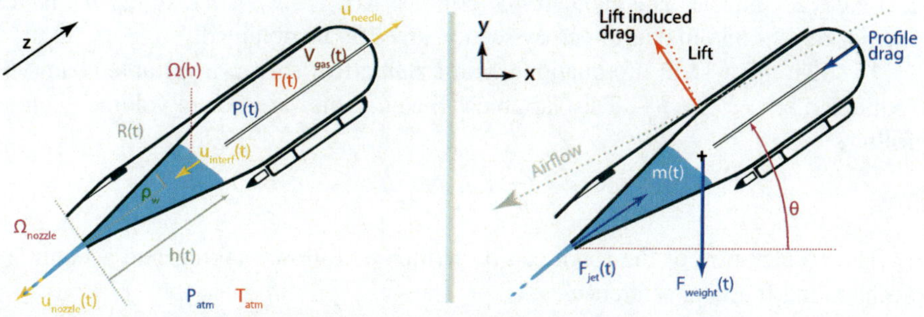

Fig. 9.4 (left) Physics consideration of the behavior of the flow in the chamber. (right) forces and definitions considered in the system

with

$$\begin{cases} T &= \rho\frac{1}{A_o}q^2 + \rho_w q\dot{Z} \\ W &= \left(m_{empty} + \rho_w V_w\right) g \\ D &= 0.5C_A A_{max}\rho_a \left(\frac{dZ}{dt}\right)^2 \\ V_w &= V_{tot} - V^* \\ g_{rel} &= g - \ddot{Z} \end{cases}$$

where we consider a 2 degree of freedom system under the influence of jetting thrust $F_{jet}(t)$ tangent to the trajectory, varying weight W, as well as drag and lift D, L from the body and the wings. The robot is at this step considered a variable mass object $m(t)$ and $F_{weight}(t)$ travelling in 2D with an initial angle of Θ. The moment M applied on the robot during jetting is approximated to 0 as the jetting force is aligned with the center of mass and the other forces (e.g. lift) are small because of the flat elevator.

Internal considerations The thrust is given by Euler's flow equation, applied in the chamber from the high pressure air zone (interface) to the nozzle.

$$\int_{nozzle}^{interface} \frac{\partial u}{\partial t} ds + \frac{\Delta P}{\rho_w} + \frac{1}{2}(u_{nozzle}^2 - u_{interface}^2) = 0 \tag{9.4}$$

Before completely leaving the water, the robot travels at a low velocity and only a fraction of its body is immersed, thus water drag is neglected. Surface tension can also be omitted at the scale of the system (≈10 cm) as it is three orders of magnitude smaller than the robot's weight.The robot's chamber is approximated by an axis-symmetric geometry, consisting of a spherical top, followed by a straight section and a conical 20° nozzle. The geometry determines $R(h)$, $A_{section}(h)$, $V_{cumul}(h)$, hence the nozzle size and maximum cross-section area can be obtained.

To solve the system of equations with a numerical solver, a variable change is required, i.e. $q = A(h)\frac{dh}{dt}$. This variable is linked to the variation of volume of air as follows.

$$\frac{d}{dt}V_{air} = -q$$

The acceleration of the robot can be written as follows, taking into account an accelerating frame of reference.

$$\ddot{Z} = \left(\rho_w \left(A(h)q^2 - \rho_w h \cdot d(q)\right) + F_{jet} + W + D\right) * \frac{1}{m(t)} \tag{9.5}$$

with $m(t) = m_{empty} + \rho_w * Vw$, where ρ_w is water density. This equation is solved for both the vertical and the horizontal axis.

Moist adiabats and non-flat water surface impact the simulation result by less than 1 [132]. These improvements are therefore neglected and the pressure ΔP is assumed to follow a dry adiabatic expansion. The initial pressure P_{init} is given by stoichiometric energy conversion, $E_{th} = m_{C_2H_2,0} \cdot HHV_{C_2H_2}$, assuming the ideal gas law holds. Flow losses through the 0.5 mm needle (u_{needle}) are neglected in this approach.

$$\Delta P = P_{init}\left(\frac{V_{init}}{V}^{\gamma}\right) - P_{air} \tag{9.6}$$

with pressure P, volume V and atmospheric pressure P_{air}. The temperature $T(t)$ can be calculated through the ideal gas law. The flow velocity u is defined at all times and points in the chamber as:

$$u(z,r,t) = \frac{dh}{dt}\left(\frac{\Omega(h)}{\Omega_{nozzle}}\right) f\left(\frac{r}{R(z)}\right) \tag{9.7}$$

The velocity is given as a function of the water level height h, the area of the interface $\Omega(h)$, the distance to the axis r and the dimension of the chamber $R(z)$.

This gives Eq. 9.8 when the variable q is isolated:

$$\dot{q} = \frac{1}{-I(h)} * 0.5\left(1 - \left(\frac{A(h)}{A_o}\right)^2\right)\frac{1}{A_h^2}q^2 + \frac{\Delta P}{\rho_w} + g_{rel}h \tag{9.8}$$

With $I(h)$ an inertial term calculated from the geometry as $\int \frac{1}{A}dz$. The system is launched at the angle given from the buoyancy calculation. During the jet phase, the jet and robot angle is considered equal to the launch angle Θ_0.

Wing lift and drag are added into the system as soon as the flap is angled, which is set to happen at the apex of the flight, i.e. when $\dot{Z}_y < 0$. The robot is assumed to immediately take on a fixed angle of attack α. The flight dynamics are given by:

$$\begin{aligned} L &= 0.5C_{L,8} * A_{wing} * \rho_{air} * |\dot{Z}|^2 \\ D &= 0.5C_{D,8} * A_{cross-section} * \rho_{air} * |\dot{Z}|^2 \end{aligned} \tag{9.9}$$

In this case study, the coefficients $C_{D,8}$ and $L_{L,8}$ were taken from CFD analysis.

9.2.2 Model-Based Design

In order to maximise the jump-gliding performance, we iterate on the analytic model of the full flight presented above, varying the nozzle diameter, volume fraction of water and launch angle. Several assumptions are taken in order to limit the parameter space. The system is constrained to a 500 mL combustion chamber as this volume corresponds to a stoichiometric mix of acetylene and oxygen using a water drop of 30 mg, or 3 times the minimum water drop volume that can be produced on-board. While the theoretical ideal combustion pressure of acetylene could reach 13 bars, static experiments show a peak gauge pressure of around 7.2 bars. We consider this

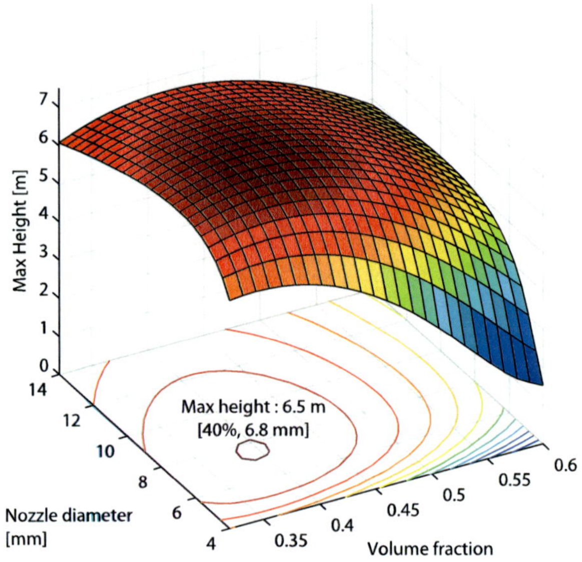

Fig. 9.5 Surface plot of analytical model output showing the effect of nozzle diameter and volume fraction on the height achieved by the robot when launched vertically

our reference flight pressure and base our optimisation on that value. The nozzle diameter can be cut to any diameter after fabrication. A larger nozzle results in a higher initial thrust, faster depletion of the chamber and consequently, faster mass loss. The initial fraction of water in the chamber has numerous implications. With more water, the weight of the robot is increased but the thrusting phase can last longer. Moreover, the initial water level also changes the floating angle and consequently the initial jetting angle.

The most important overall performance metric for the robot is the distance travelled per jetting cycle, which potentially maximises the usefulness of its mission. The results of the optimisation for maximum height carried out with the physics model are shown in Fig. 9.5. The maximum flying height is achieved with a nozzle of 7.3 mm and a water fraction of 39%. However, one can see that the flight height (and distance) does not depend significantly on nozzle size or volume fraction within a large range of these parameters. A variation of ± 1 mm and $\pm 10\%$ water fill changes the total distance by only 3 m, or less than 10%. This gives a wide design space without major impact on performance. The nozzle is cut to a diameter of 6.8 mm, yielding a final total volume of 491 mL and a water fraction of 40%, corresponding to 196.4 mL. Finally, the optimal launch angle for maximum flight distance is found analytically to be 47° and the mass distribution of the robot is altered so that it floats at that set angle.

Numerical simulation of jetting phase A 3D computational fluid dynamics study of the flow inside the chamber in the propulsion phase was carried out to compare the results obtained in the physics model and better understand the flow behaviour in 3D, ensuring that the analytic model represents conditions as realistically as possible before manufacturing the robot. The simulation provides insight into the effect of

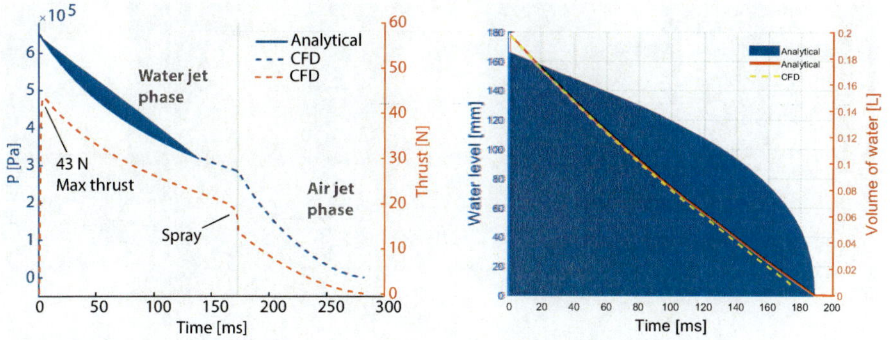

Fig. 9.6 Pressure, Thrust, water level and volume evolution

friction of the fluid against the internal walls of the chamber, the thrust produced by the jetting air once the water is evacuated, the irregularity of the water-air interface during jetting and the effect of water-air spray caused by fluid instabilities at the end of the water jetting phase. The average reduction in thrust due to shear in the inner walls was shown to be a negligible 0.39% in the simulation. Figure 9.6 (right) shows that the 2D analytic model and 3D numerical simulation agree in their prediction of the variation of water volume with time, and Fig. 9.6 (left) that they agree with the evolution of pressure during water jetting. The pressure in the chamber decreases from 6.5 bar to 2.84 bar during the water jetting phase, while the thrust reaches a maximum of 43 N and decays to 20 N after 161 ms, when the jet becomes a mixture of air and water. As a higher proportion of air starts to be ejected, the thrust reduces further to 14 N at 173 ms, when the jet is made up of only air. Due to the lower density of air, the pressure and thrust decrease more rapidly during this phase, reaching 0 at 285 ms.

9.2.3 Static Modelling

The performance of the propulsion system hinges on various geometric parameters. Total thrust is a key performance metric, which is determined experimentally on a static 1:1 model of the combustion chamber of the robot. This transparent model, shown in Fig. 9.7, is also a platform for video analysis, as both the acetylene production and combustion can be viewed. For the purpose of analysing the motion and initial stages of jetting, the model is mounted in a tank with a transparent side.

Notably, slow-motion analysis of the combustion process shows that the speed of progression of the flame far exceeds the speed of the water front. The inertia of the water contained in the bottom of the chamber is such that the pressure can fully build up the top of the chamber before the volume changes significantly. This allows for simplification of the physics model. As such, the dynamics of the water flow can be solved with an initial high pressure in the chamber, considered to have happened

Fig. 9.7 Propagation of the flame front

instantaneously. The combustion propagates spherically, starting from the electrode tips until it reaches the side walls. It then continues to expand until the combustion is complete when the front has reached the corners of the chamber (Fig. 9.7 right).

Igniting a gas in a wet environment at atmospheric pressure is challenging. The transparent test chamber permitted testing of various parameters of the electrodes. It was found that the material of the electrode tips has a limited influence on the likelihood of generation of an electric arc. Copper, brass and zinc-plated steel were successful in generating an arc. Visible damage of the electrodes was noticeable after multiple arcs, and was more pronounced with brass and copper. However, even with strong degradation, the feedback loop of the sparking circuit increased the voltage over the breakdown threshold. The distance between the electrodes plays a significant role. If the electrodes are too far apart, the arc does not occur or, worse, might happen somewhere else in the circuit. Conversely, if they are too close, the gas does not ignite even if the arc is formed. The gas mix is also important as a richer mix is easier to ignite.

The transparent test chamber is designed to have the same geometry as the flying prototype in order to closely match its jetting characteristics. Both the force and the pressure evolution are measured as the robot is attached to a rigid frame. The water level evolution is simulated numerically and is found to closely agree with the analytically computed water level, as shown in Fig. 9.6 right.

The static thruster's structure consists of a transparent tube with two inserts, i.e. a spherical top and a conical nozzle bottom. The tube is overdesigned, with walls of 6 mm thickness. This guarantees the safety of the experiment, permits close monitoring and does not negatively impact performance as this model does not need to fly. The acetylene combustion reaction happens in the middle section, where the adjustable ignition electrodes are placed on the right side, as well as a metal gas injection tube and a pressure sensor output. The electrodes generate an electric arc from an oscillating high-voltage source. The pressure is measured using a gauge sensor read by an acquisition card at 5 kHz. The whole structure is affixed to a force sensor. This setup permits direct comparison of the pressure to the resulting force in a static frame.

9.3 Design of an Aquatic Jump-Glider

Based on the knowledge acquired through modelling and static tests, a new robot design is developed. The flight-capable thruster development is presented in this section, with a discussion of the operating principles and the hardware and software implementation.

9.3.1 Robot Design

The successful implementation of an aerial-aquatic robot that employs combustible gas obtained on-board for impulsive take-off poses significant challenges.

Firstly, the management of the generated high pressures on a small robot imposes stress on its structure and hydraulic system. Achieving consistent combustion inside a flooded chamber is also not a straightforward process due to the influence of water on ignition and combustion.

Moreover, the robot needs to operate in 3 distinct modes; jetting, gliding and water surface operation, which includes landing, data collection and refilling Fig. 9.2. Each of these stages has significantly different and sometimes conflicting requirements, posing one of the greatest challenges to the design of such a robotic platform. This leads to trade-offs in the design stage, which limit the performance achievable in each locomotion mode.

For the vehicle to fly in a stable manner during the jetting phase, the center of mass must be a significant distance in front of the center of pressure of the vehicle. However, to maintain a stable floating position on the water surface and the desired angle during jetting, the center of mass must be located behind the center of buoyancy. For the gliding phase, a fine balance between the center of mass and center of pressure must be struck in order to achieve static longitudinal flight stability passively. During gliding, the center of mass should be slightly forward with respect to the wing's center of pressure. This, however, creates a nose-down pitching moment, which is countered in this robot by an elevator providing a counter nose-up pitching moment. Further considerations that affect the vehicle configuration are the performance of the propulsion system, the vehicle aerodynamics and the change in the center of mass during jetting.

Using the aforementioned principles, this section describes a 160 g robot capable of exiting the water in an uncontrolled environment by reacting 53 mg of calcium carbide with the surrounding water. The reaction generates an average peak pressure of 6.5 bar in 25 ms, which accelerates the robot to 10 m/s and yields a flight distance of 26 m. The pressure, measured on-board during jetting, sharply increases up to 7.8

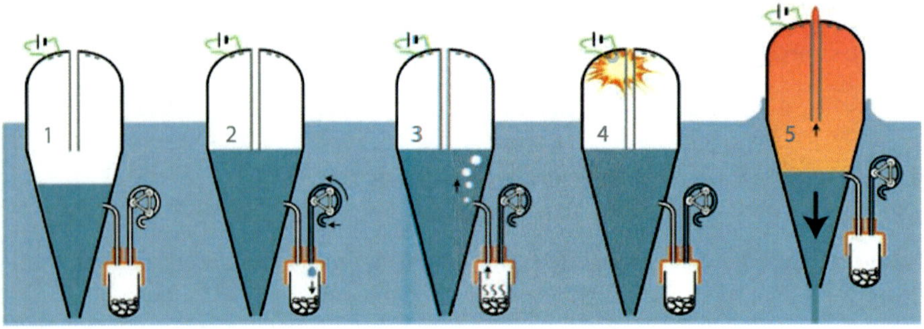

Fig. 9.8 Illustration of the 5 reaction phases. 1. Water intake 2. Water pumping into fuel tank 3. Reaction to produce acetylene gas 4. Ignition 5. Water jetting, showing choked needle

bars in 25 ms. The system empties in about 200 ms, after which it transitions to a glide before landing back in the water.

9.3.2 Principle of Operation

The robot creates acetylene gas from the reaction of calcium carbide with water, after which the ignition is triggered Fig. 9.8. As soon as the robot lands on the water it starts filling up to the right level due to its low center of mass and venting needle (phase 1). The length of the needle limits how much water can flood the chamber and consequently guarantees that the robot stays afloat at a predefined waterline. Maintaining this preset water line reduces water drag compared to take-off from a completely submerged state. In addition, wireless communication is guaranteed, something not possible during submersion. The micro peristaltic pump is then turned on and produces one droplet every 3.3 s, resulting in a rich gas mix in about 10 s (phase 2). The droplets react with the fuel quickly and the resulting acetylene gas escapes into the combustion chamber (phase 3) as the pump blocks the other opening. The gas is then ignited with a high-voltage electric arc (phase 4). Thrust is generated as soon as the pressure increases (phase 5).

Figure 9.9 shows a bottom view of the robot. The driving components fit under a sealed transparent dome, affixed to the composite combustion chamber. The multi-launch fuel container rests at the interface to permit reloading without the need to open the dome. A high delta wing, elevator and vertical wingtip fins are attached on the opposite side of the chamber.

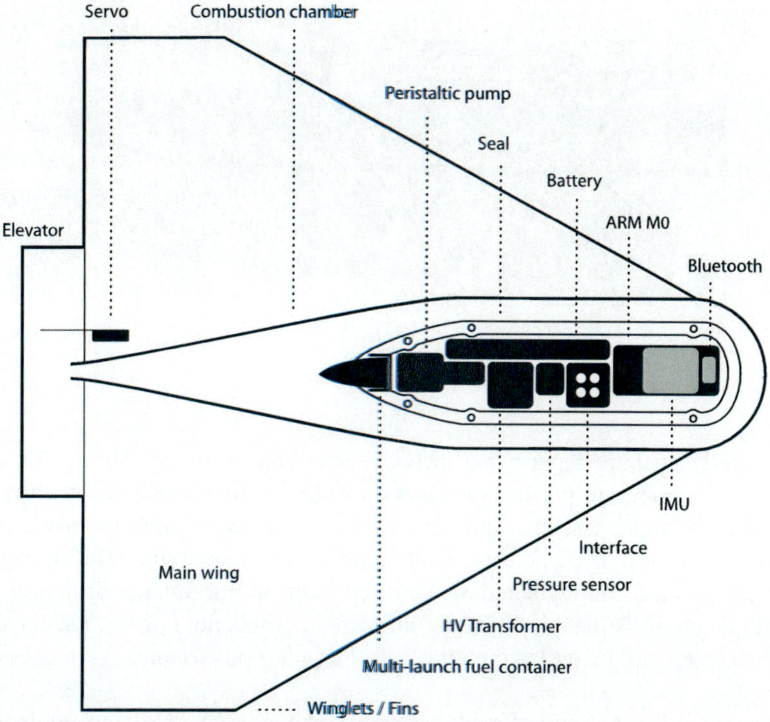

Fig. 9.9 Bottom view schematics of the aircraft with the electronics on top of the combustion chamber

9.3.3 Systems

Combustion Chamber The combustion chamber is designed with a rounded top followed by a straight cylindrical section, a 20° conical section ending on a nozzle as shown in the side view (Fig. 9.10). The chamber is designed to withstand 8 bars of internal pressure, following the results from the static experiments.

Driver module Permanently bonded with flexible epoxy to the combustion chamber is a 3D-printed nylon frame. This part not only holds every individual component but its edge also supports a seal. Finally, there are six screw inserts in the frame to attach the top dome. This dome is a vacuum formed high-impact polystyrene 0.75 mm transparent shell that is screwed to the frame, compressing the laser-cut 1 mm silicone seal. The whole assembly is fully submersible and visible due to its transparency. The calcium carbide tank consists of a 1 mL chemical test tube, with one teflon tube for water inlet and one PVC tube for gas outlet, both through the cap of said tank. This cap is bonded to the dome.

Under the dome, a small rectangular removable plate interfaces the electronics module with the combustion chamber. This device increases the lifetime of the robot

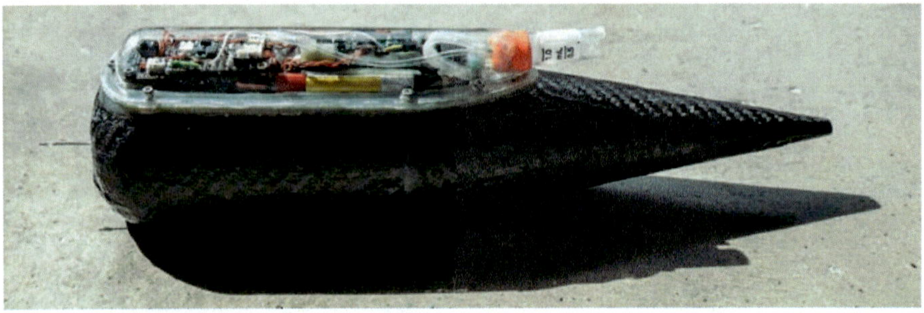

Fig. 9.10 Side view photograph of the assembled jetting system showing the electronics under their transparent waterproof cover

significantly, making repairs and inspections straight-forward. The sealed carbon fiber plate is bolted to pre-installed screws welded to the inside of the chamber. It bears the acetylene gas inlet, the pressure tube connection and two insulated high voltage electrodes. The micro peristaltic pump's DC motor (RP-Q1 Takasago Fluidic Systems) runs on regulated 3.3 V for constant speed. Smaller droplets provide a better count of the amount of water dispersed per second. For that reason a teflon tube is attached to the outlet, creating 3.3 ml droplets per second.

Electronics The waterproof module described above contains all the required flight electronics. At its centre, a micro-controller is connected to its actuators (Fig. 9.11). The water droplets are created by a peristaltic pump, which draws water from the surrounding environment. This pump has proven to be extremely reliable in real world conditions. A gauge pressure sensor measures the chamber pressure after ignition. Lastly, the system requires an ignition source. The consistent ignition of the acetylene is a challenge at small scales, especially at ambient pressure and with a half-flooded chamber. Hydrogen ignition at smaller scales has been demonstrated [179], however with the help of an off-board power supply. Previous research generated a spark from a piezoelectric element compressed by a shape-memory alloy wire [184]. This mechanically actuated system wears out quickly due to the high forces in play. This robot solves the ignition issue with solid state electronics that consist of a feedback transformer and a power transistor to create a pulsed electric arc of 3 mm.

The robot has an onboard 2.4 GHz communication link running the Bluetooth Low Energy 4.0 protocol. The electronic capsule is arranged so that the antenna faces the top of the robot, or else the signal gets weakened by absorption in the water. The bluetooth link is set up as a UART tunnel, transmitting serial data directly to a central mode device such as a phone. The control of the launch happens in three distinct phases. Firstly, default flight parameters can be changed before launch, e.g. water amount reacted. Secondly, launch is initiated. After the experiment is over, recorded data is streamed back to the connected device.

The whole system is powered by a 270 mAh lithium battery.

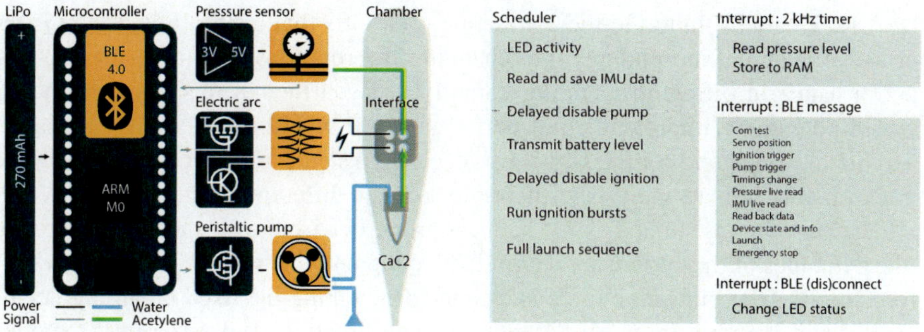

Fig. 9.11 (left) System diagram of the on-board electronics and transducers. (right) embedded code execution

Embedded software and control The robot is controlled by a microcontroller board,[1] which combines an ARM M0 processor with a bluetooth low-energy chip, shown in Fig. 9.11 left. At 5.8 g, this board is a good trade-off between low weight, sufficient number of inputs and wireless communication. Most onboard functions are executed at determined times. Simultaneously, the robot captures inertial and pressure data (Fig. 9.11 right). The IMU sensor fusion is realised on chip and the resulting data recorded 100 Hz. The gauge pressure sensor is read at a 2 kHz rate and provides accurate values despite the harsh conditions. Bluetooth communication and telemetry commands are also handled meanwhile.

Wing Fabrication The wing is cut from of a 0.2 mm thick flat carbon fiber plate with fibers in a 2/2 twill weave oriented span-wise and chord-wise. A high wing configuration is used to ensure roll stability and two vertical stabilizers are located on the wingtips to reduce induced drag. Pitch stability is ensured by performing a CFD analysis and studying the static stability of the robot in flight. The delta wing configuration results in two vortices forming along the leading edges at high angles of attack, which delay flow separation at high angles of attack and creating large suction at the top surface, enabling higher lift coefficients as a consequence. The robot is able to generate enough lift for relatively small wingspans, keeping the weight of the system to a minimum.

9.3.4 Wing Design and Stability

The multi-modal capabilities of the platform introduce conflicting requirements for its design, which are linked to the different dynamics that describe its locomotion stages. The wing's geometry was used as the last design variable to ensure good behaviour of the platform throughout its operation.

---

[1] Adafruit Feather BLE M0.

A reasonable solution to the conflicting trade-offs that control this design is the usage of adaptive morphology, i.e. changing the robot's shape to ensure the best performance in different operating conditions. Even though this design principle would potentially improve overall performance, it would do so at the expense of additional weight and complexity. It was decided, instead, to balance the wing design trade-offs through an analysis of their effect on the different mission phases.

The buoyancy is studied using the mass properties from a 3D model of the robot. The dynamics of the system are simplified, taking only the weight and the buoyancy force in 2D. The buoyancy force is estimated by taking the fixed breathing needle length, and iterating immersion depths (h) and flotation angles (α) in the model. This permits the calculation of the total water volume displaced and consequently the buoyancy force and center of buoyancy. These are evaluated in order to identify the static stable equilibrium configuration of the system in terms of h and α. This equilibrium configuration corresponds to the h and α values for which the value of the buoyancy force equals the weight and the center of buoyancy of the robot produces no moment around the center of mass. Moreover, the stability condition during floating is that the center of buoyancy is located above the center of mass.

The wing is thus designed as a denser than water, yet light, thin plate with the center of mass located towards the back of the robot. An uncambered thin plate is used as it generates no lift for 0° angle of attack, and only skin friction drag during jetting at the air-water interface. Half circle unidirectional carbon fiber rods are used as stiffeners to increase stiffness and ensure stability in flight.

A delta wing geometry is used as it delays stall when compared with a regular rectangular wing and because its center of pressure varies with the angle of attack, positively impacting the robot's stability during both flight stages. This property allows the robot to keep its center of pressure behind the center of mass for any volume of water inside the chamber while also maintaining a static margin during gliding that allows reasonable maneuverability. An elevon, which counteracts the moment generated by the lift and drag is added to the back of the wing. Its angle is set to a fixed value so that gliding will occur at 8° the sustained gliding angle at 10 m/s.

Numerical simulations are performed in order to calculate the wing area necessary to achieve sustained gliding, and to ensure pitch stability of the robot during gliding and jetting. The static stability of the robot in flight is ensured by obtaining negative derivative of the pitching moment with the angle of attack, and the calculated sustained gliding angle is obtained as 8° for a velocity of 10 m/s, which corresponds to the maximum glide ratio of the robot.

9.4 Cyclic Aquatic Escape with Jet Propulsion

The water-reactive fuel thruster, integrated into a fully functional flying vehicle was extensively tested. Over a total of 22 flights and multiple prototypes, the vehicle was

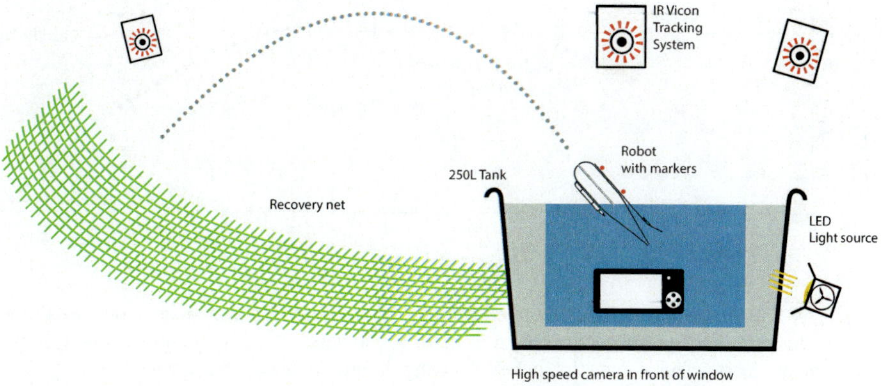

Fig. 9.12 Labsetup

operated in controlled laboratory conditions, in waves and in uncontrolled, windy, outdoor from stagnant ponds.

9.4.1 Controlled Conditions

Most buoyancy, refilling and jetting tests were carried out in the lab, shown in Fig. 9.15. The jetting phase can be accurately studied as well as the initial part of the trajectory. A 250-liter tank was modified with a PMMA window in the upper section of the side panel, shown in blue in Fig. 9.12. The acrylic panel is 6 cm larger than the cutout in the tank and securely bolted in place on the inner side of the tank. A high-speed camera[2] is placed at water level height, recording half of the underwater scene and half of the air scene. In addition, a 150 W LED lamp specially engineered for this experiment lights up the underwater scene to allow up to approx. 2500 fps filming. The tank is setup in the far side of the flight arena, a 12 m × 6 m space with a 6 m ceiling. Suspended nets reach from the walls down to the tank to protect the robot from ground and wall impacts.

The whole space is tracked by the 16-camera 3D Vicon tracking system for position and velocity measures. The grey circles visible on the fins are 2D reflective markers for the tracking system. For better tracking, spherical reflectors (shown in Fig. 9.12) can also be employed although they do negatively impact the aerodynamics of the vehicle.

Lastly, two high-resolution cameras are set to record the jetting phase, placed in two corners of the flight arena

The robot will upright itself to the designed angle of 47° when it lands in the water due to its center of gravity being behind the center of buoyancy (Fig. 9.14 right). As soon as it rests on the water surface, the nozzle is underwater and the

[2] TS5Q Fastec.

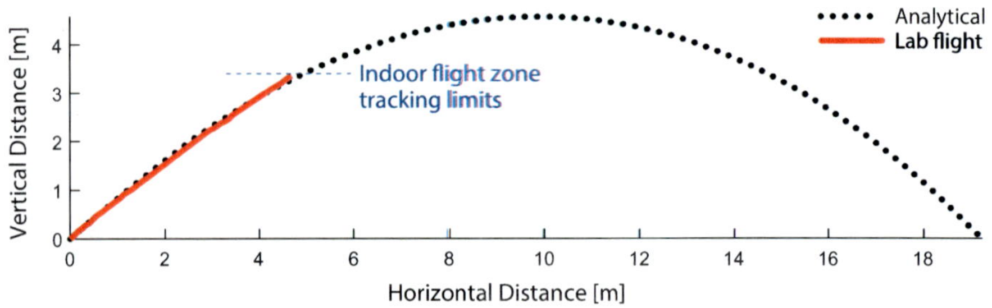

Fig. 9.13 Indoor flight trajectory obtained from the motion tracking system and compared to the modeled trajectory. The results are found to be in accordance although we note a slight drop in experimental performance due to chamber and wing deformation and pressure losses

chamber starts to fill up. The reader can note that the refilling time is significant. This is a design decision that only necessitates a static element in order to vent the extra air, fix the water level and is insensitive to the harsh test conditions. For all of the characterisation and flight tests presented here, the refilling time is irrelevant and as such, the method adequate. The venting needle could be replaced with a valve for faster reloading times.

The time spent refilling could be spent on the robot's main mission, taking water measurements or sampling data. Once the orientation of the robot matches the expected angle, the launch sequence is triggered, and the robot takes off.

The velocity of the robot during the initial stages of flight is compared between the analytical model output and the tracking system measurements (Fig. 9.14 left). The tracked velocity can be seen to follow the modelled velocity closely, being slightly slower at the end of the jetting and the beginning of gliding. This is linked to chamber deformation under high pressures, wing bending which has a major impact on flight angle, pressure losses and variability in pre-combustion gas mixing. The displayed velocity is obtained from the tracked trajectory shown in Fig. 9.13. The robot's trajectory is shorter than modelled due to the slightly lower velocity achieved at the end of the jet.

The corresponding accelerations are displayed in Fig. 9.14 (tracked and predicted) and agree well, albeit with a sharper drop at the end of the jet in the model which does not consider air thrust. Inertial measurement IMU from three separate flights show acceleration profiles in agreement with tracking data, with some variability stemming from launch angle inaccuracy and jetting differences between experiments. The onboard sensor saturates at 4 g as indicated by the dotted black line and samples at 100 Hz.

Note that in the model, the jetting stops earlier and more abruptly than in the experiments, at about 200 ms, since it doesn't take into account the spraying or air jetting as described in the Physics of jet-gliding section above. The measured duration of jetting ends at 285 ms, which is consistent with the CFD simulations. Tracking stops just after 600 ms when the robot leaves the field of view of the cameras.

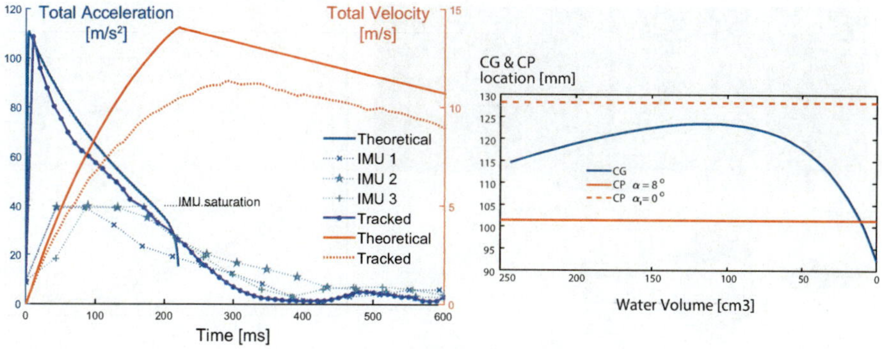

Fig. 9.14 (left) Comparison of position and velocity profiles for the experiment and the analytical model. (right) Evolution of the position of the center of gravity as the robot empties, moving ahead of the center of lift

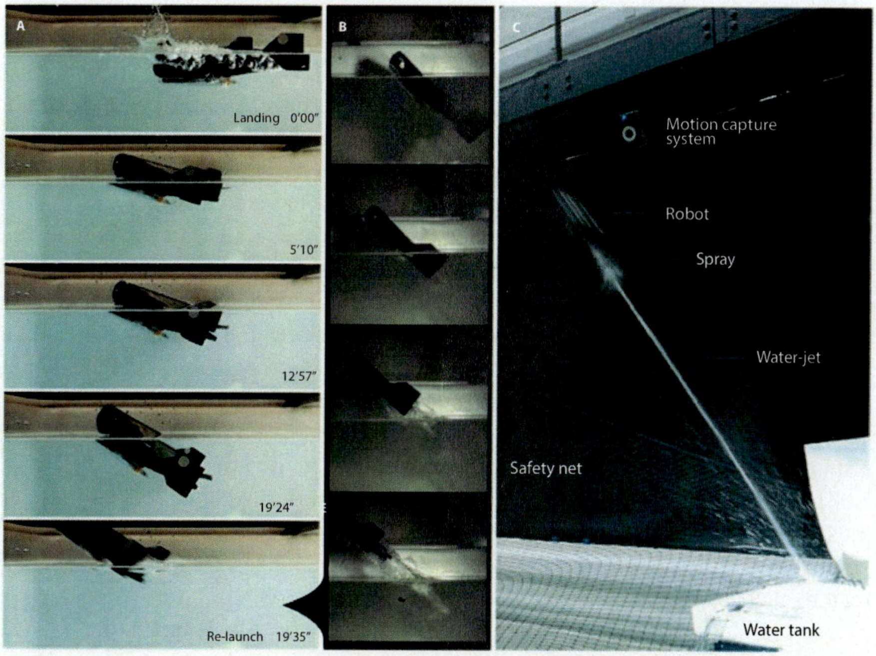

Fig. 9.15 Time-lapse of the landing process, refilling process and subsequent launch. (**B**) Capture of a high-powered launch in the flight arena at the end of the water jetting phase. (**C**) Robot during water escape

9.4.2 Outdoor Conditions

Outdoor flight tests validate the reliability of the robot in uncontrolled conditions. Indeed, the water composition, impurities, temperature, wind, water surface state and

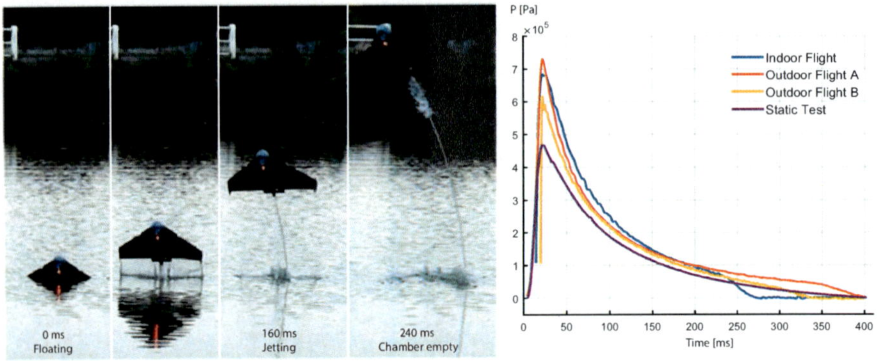

Fig. 9.16 (Left) Composite image of the robot in floating, jetting and flying mode. *Bystander on the side-walk edited out for clarity. (Right) Pressure evolution of indoor, static and outdoor flights, aligned at pressure peak

more factors are varying and unpredictable. In addition, the distance to the lab vastly reduces the ability to alter or repair a vehicle to achieve a successful experiment. Outdoor tests are a vital step towards full autonomous deployment and prove that the robot can function outside of an artificially arranged set of specific condition.

Outdoor water exit is recorded from a pond which has a depth of about 1 m at the launch position and is subject to outdoor, gusty conditions. The robot is oriented in a specific direction for video capture and is loosely kept in place by two rods to avoid drifting. The control commands are sent from shore. Figure 9.16 (left) shows the robot flight during four key phases of the water take-off. In the first stage, it fills up with water while venting air through the top needle. As the pressure in the chamber increases, thrust is generated and the robot exits the water. The jetting continues in air during the first 200 ms of flight. Once the chamber is empty, spray is visible for a short period until the pressure equalizes as shown in the last time shot.

Three different pressure measurements are taken during flight by the on-board pressure sensor and compared in Fig. 9.16 (right). The on-board recorded pressure is validated with the off-board sensor installed during static tests, shown in Sect. 9.2.3, read by means of an acquisition board and Labview.

The pressure peaks achieved in flight vary between tests by a maximum of ±15% of the peak pressure, consistent with different fuel mixes. The exact gas mix ratio depends on the dispersion of water droplets onto the calcium carbide, which varies with the tube surface tension, orientation of the robot and vibrations. In addition, visual recording of the combustion process, as can be seen in Fig. 9.3 shows non-spherical evolution of the flame front. Multiple experiments reveals variation of the shape and evolution of the combustion fronts, which has an impact on the recorded pressure peak. Lastly, the peak might reach higher values between measurements as the sensor operates at 2 kHz. There is more variation with the static experiment, linked to pressure losses in the different construction.

Fig. 9.17 Flight trajectory of a pond-to-grass launch

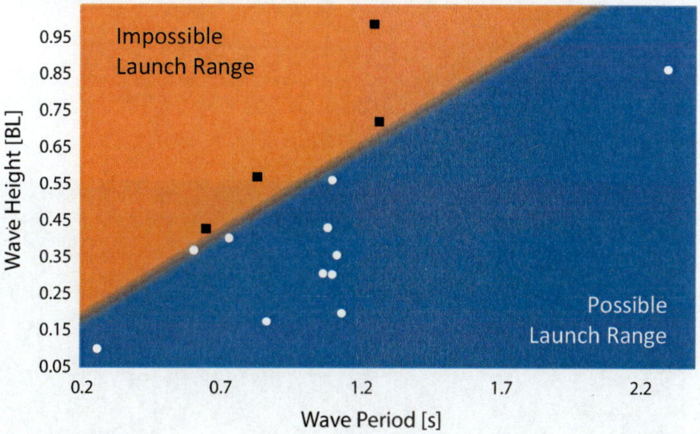

Fig. 9.18 Launch capability of the aerial-aquatic robot in waves of varying period and height (trough to crest). The robot can withstand big waves as long the wave period is sufficient. The experimental tank used limits the maximum wave height to the robot's size, i.e. 30 cm from crest to trough

Wind was recorded to be an average of 5 m/s with gusts of 7.5 m/s. The gusty wind conditions create variance in the overall flight distance, however successful take-off was possible with changing wind speed and direction. Figure 9.17 presents a flight trajectory. Here, the robot exits the water level to land on grass. By placing the robot back into the pond, it performs multiple flights.

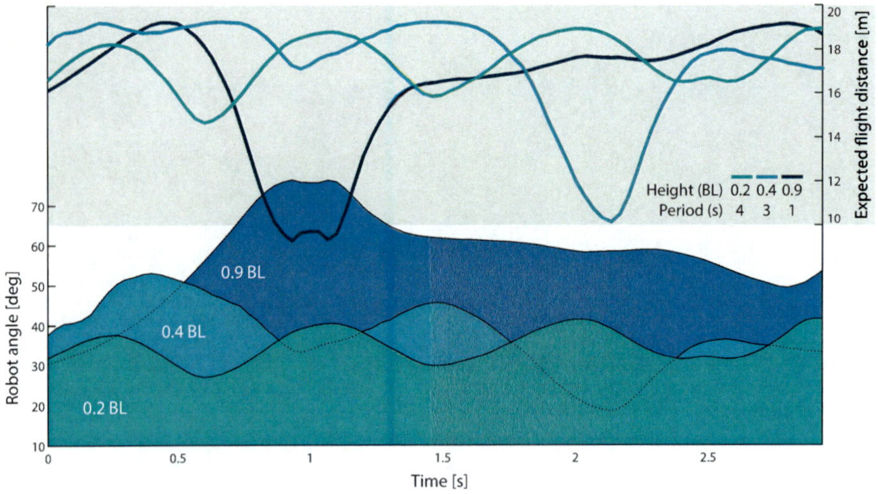

Fig. 9.19 Instantaneous impact of 3 differently sized waves on the orientation of the robot as well as the expected flight distance at each point in the wave. The plot shows 1, 3 and 4 periods of a big, medium and small sized wave (0.9, 0.4 and 0.2 body lengths)

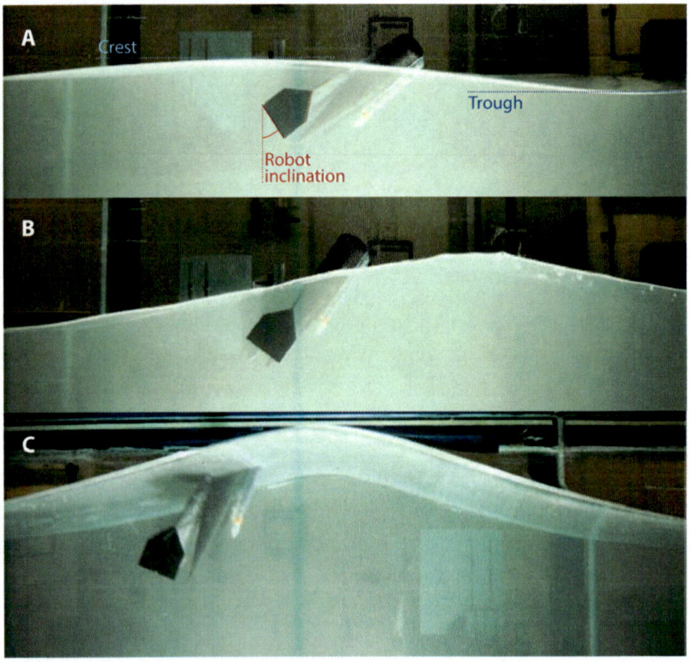

Fig. 9.20 Experimental study of the robot flotation angle in various wave conditions. (**A**) 0.2 BL height, 0.9 s period (**B**) 0.4 BL height, 1.1 s period (**C**) 0.9 BL height, 2.3 s period. The angle is tracked over time at 30 fps and the data used to generate Figs. 9.18 and 9.19. The waves are produced in a 1 m wide channel with a length of 9 m

9.4.3 Wave Tolerance

The thruster presented in this research has been demonstrated to perform multiple aquatic take-offs, from a tank and from an outdoor pond. Due to its impulsive actuation force of over 25 times the robot's weight, it is relatively insensitive to the water surface state and can jet off in wavy conditions. The robot will successfully escape the water when angled between 10° and 90°, based on modeling and confirmed by experiments. The robot's rotation has been tracked experimentally for various waves sizes and periods (Fig. 9.19). If the robot is hit by a particularly steep wave from the front, which is the most critical case, it will tip over and take-off becomes impossible. This occurs for waves with a large height to period ratio, as shown in Fig. 9.18. Nonetheless, the robot exhibits a good expected flight performance at most points in different waves, as reported in (Fig. 9.20).

Chapter 10
Efficient Water-Air Propulsion with a Single Propeller

10.1 Introduction

In the previous chapters, aquatic launch and dives into water with small flying robots have been demonstrated. An AquaMAV prototype was presented which was capable of self propelled-flight in air and able to escape water, but this robot had no means of propelling itself beneath the surface. To add aquatic locomotion it is attractive to use the same propulsion system as is used for flight, as this reduces the weight and complexity of the system. However, the increase in load on the propeller in a denser fluid results in much slower rotation speeds, and means a significant loss of motor efficiency.

In this chapter, the efficiency of propeller operation in air and water will be examined again using QPROP (see Chap. 7) to investigate a variety of propeller geometries commonly used in Micro Aerial Vehicle (MAV) propulsion. These theoretical predictions are then validated against underwater thrust tests, and the model is used to investigate an ideal aerial-aquatic transmission. Based on the results of these investigations, a bimodal epicyclic transmission to allow efficient propeller operation in air and water is proposed. This gearbox is then fabricated and tested. By using the direction of rotation of the input motor to control the setting of the gearbox, the proposed design provides a simple and effective solution to achieving efficient aerial-aquatic locomotion. In Chap. 6 several existing designs for aerial-aquatic propulsion were detailed, including the use of aerial propellers underwater. The aquatic quadcopter systems [108, 109] simply force aerial propulsion systems to operate underwater, where they will be highly inefficient. The system presented in this chapter significantly improves the overall performance of propulsion in the two media, with minimal mechanical complexity and no additional actuation.

R. Zufferey et al., *Between Sea and Sky: Aerial Aquatic Locomotion in Miniature Robots*, Biosystems & Biorobotics 29, https://doi.org/10.1007/978-3-030-89575-4_10

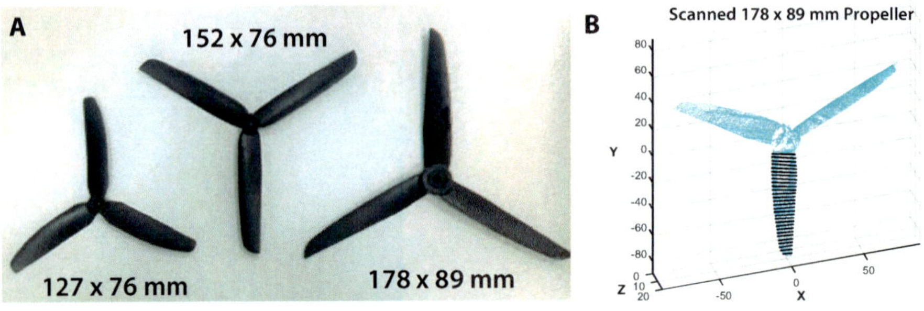

Fig. 10.1 (**A**) 3-blade variants of propellers tested in Sect. 10.3. (**B**) Point cloud from 3D scan of propeller geometry (cyan dots), with extracted blade chord lines marked in black

10.2 Computational Investigations

The differences between electric propeller operation in air and water were first investigated by analysing motor-propeller combinations using QPROP, examining the effect of propeller geometry, motor parameters and operating conditions on the thrust production and efficiency of the overall system.

As in Chap. 7, propellers are identified by their diameter and pitch (Diameter × Pitch, Fig. 10.1A). The propellers tested in Sect. 10.3 were modelled using a 3D laser scanner (FARO Edge ScanArm, 25 μm accuracy) to generate surface point clouds (Fig. 10.1B). These were then parsed into QPROP input files using Matlab. A cylinder is fitted to the central hub to locate the blade's rotation axis, and cross sections are taken of the blade scan at 30 equally spaced sections, from which chord length, pitch angle and airfoil sections can be extracted for input into QPROP.

Our objective in this investigation is to maximise the efficiency of an aerial-aquatic propeller propulsion system, achieved by matching propeller and motor efficiencies. The definitions of efficiencies used in Chap. 7 are retained, repeated here for convenience. Propeller efficiency is defined by the ratio of propulsive power out (thrust, T multiplied by forward speed, v) to shaft power in (shaft torque, Q multiplied by angular speed, ω),

$$\eta_P = Tv/Q\omega \tag{10.1}$$

and motor efficiency is the ratio of shaft power to electrical power:

$$\eta_M = Q\omega/VI \tag{10.2}$$

where V and I are the input voltage and current. Total system efficiency is then:

$$\eta_T = \eta_P\eta_M = Tv/VI = P_o/P_i \tag{10.3}$$

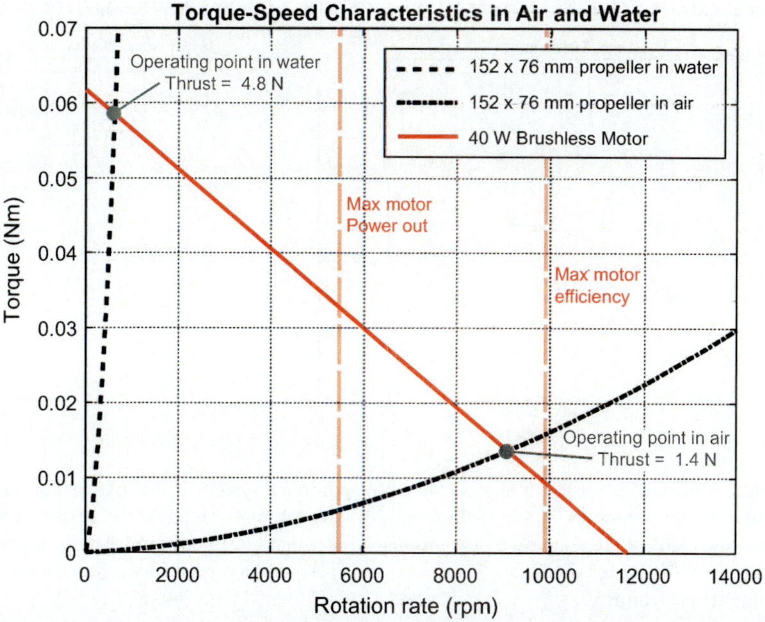

Fig. 10.2 Torque speed curves for a 152 mm diameter, 76 mm pitch propeller in air and water and a 40 W example brushless motor. The rotation rate at which the motor operates at maximum efficiency (η_M, Eq. 10.2) and produces maximum output power ($Q\omega$) have been marked with dashed lines. The torque-speed curve of the propeller shifts steeply to the left in water, increasing equilibrium torque and decreasing operating rpm. In air, the propeller operates at near maximum motor efficiency, but in water the motor is at only 6% of its peak efficiency speed

where P_i and P_o denote input and output power. Efficiency is therefore zero when static and $v = 0$, and is strongly a function of forward velocity, which determines the relative motion of propeller blades to the surrounding fluid.

However, to highlight the problems of operating aerial propellers underwater, the torque requirements of a static 152×76 mm propeller in air and water have first been analysed, when driven by the brushless motor currently used for aerial propulsion in the AquaMAV described in Sect. 7, and again using a first order motor model (Sect. 7.2.2, Eq. 7.4). This motor has a peak output power of 40 W. Figure 10.2 shows the torque required to drive the propeller against rotation speed in water and in air, and the torque-speed characteristic of the AquaMAV motor. The rotation speed at which a motor produces maximum output power, ($Q\omega$) and maximum efficiency (η_m, Eq. 10.2) can be calculated analytically, and are marked on the plot. When the motor output is connected to a propeller, the operating point of the motor-propeller system is at the intersection of the motor and propeller characteristics (Fig. 10.2). Here it can be seen that the increased fluid density in water shifts the propeller characteristic backward significantly, forcing the system to operate at a much lower rpm. While the torque is increased, the reduction in speed means that the two propeller flows are dynamically similar, and blade tip Reynolds numbers,

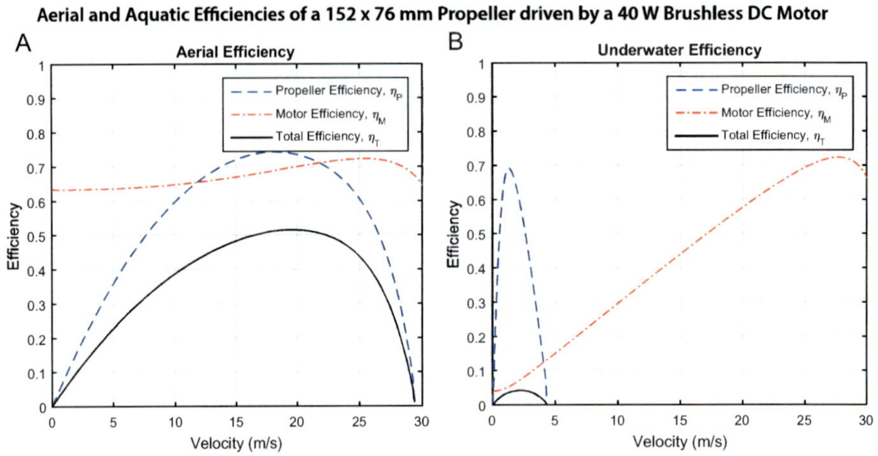

Fig. 10.3 Efficiencies curves of the same motor-propeller system in air (**A**) and in water (**B**) showing that the properties of the medium greatly affect the matching of the system. Efficiency curves shown are for the coupled motor-propeller, such that the forces produced by the propeller effect the operating speed of the motor and *vice versa*. Beyond a certain forward speed $\eta_P < 0$ and the propeller is creating drag

$$Re = \rho\omega D^2/\mu \tag{10.4}$$

where ρ is the fluid density, μ viscosity and D propeller diameter, are 1.08×10^6 and 1.20×10^6 in air and water respectively. However, the thrust produced in air is 1.4 N, while over three times the thrust is produced in water (4.8 N). So while the motor can produce significantly more force underwater, it must do so at 10% of its maximum power output speed, and 6% of its maximum efficiency speed.

The matching problem of aerial aquatic operation can be seen more clearly when forward velocity is taken into account. In Fig. 10.3A, B the efficiencies of the same motor-propeller combination are plotted based on QPROP output data over a range of forward velocities in air and water. Beyond a certain speed, the propeller cannot rotate quickly enough, the propeller creates drag rather than positive thrust, and $\eta_P < 0$. It can be seen that while maximal propeller efficiency is not greatly reduced in water ($\eta_P = 69\%$ compared to 75% in air), the motor speed is much lower, forcing the system to operate at a very low efficiency ($\eta_T = 5\%$ compared to 51% in air). This highlights the key problem in using an aerial propulsion system underwater; it is difficult to achieve good motor-propeller matching in both media. Conversely, if the motor used has a torque characteristic appropriate to the higher resistance in water, the low maximum rpm will result in negligible thrust in air.

The curves shown in Fig. 10.3A, B show the motor operating at full power. When operating at lower power, the efficiency of the system is expected to increase. This is because the equilibrium torque of the system increases with the voltage applied, increasing the mismatch between the motor design torque and required torque. In Fig. 10.4, the effect of changing the motor supply voltage on efficiency is shown. The

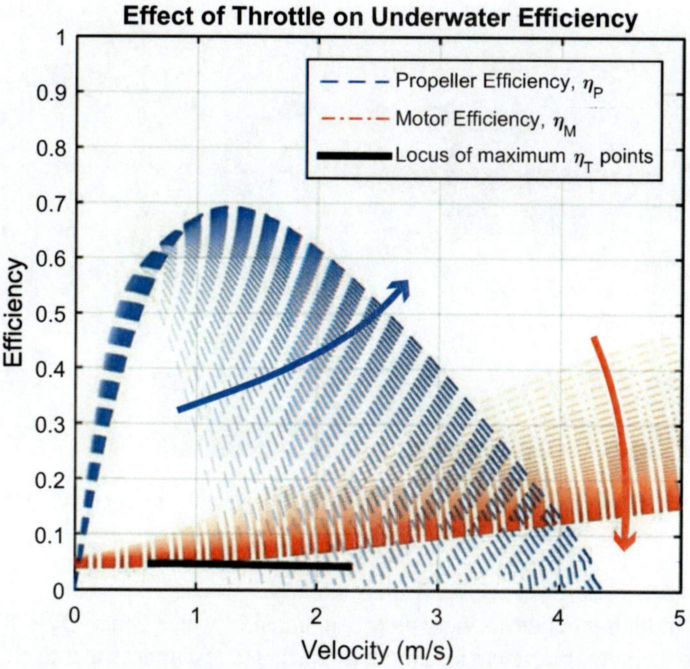

Fig. 10.4 The effect of throttling the motor on underwater efficiency by increasing the applied voltage from 2 V to 8 V. The locus of maximum total efficiency points is marked, showing that throttling has little effect on total efficiency for a mismatched system. Efficiency curves at lower voltages are shown with increased transparency

motor is simulated operating from 2 V to 8 V, and the maximum total efficiency at each voltage is calculated. The locus of these points is plotted in Fig. 10.4. Propeller and motor efficiency curves are also shown for each simulated voltage, plotted with decreasing transparency as motor voltage is increased. The results show a small but insignificant increase in total efficiency as voltage is reduced (from 4.2% at 2 V to 4.6% at 8 V). This shows that although the vehicle will not necessarily always operate at full throttle when travelling, reducing motor power has little effect on the maximum achievable total efficiency.

10.3 Experimental Verification

In order to verify the computational predictions made using QPROP, several propellers were tested underwater (examples shown in Fig. 10.1A) and torque-speed curves were recorded, using encoders (Pololu 12CPR Magnetic Encoder) and a 6-axis force balance (ATI Gamma IP68). The qprop code had previously been tested against experimental data in air (Fig. 7.3), and since the flows are at similar Reynolds

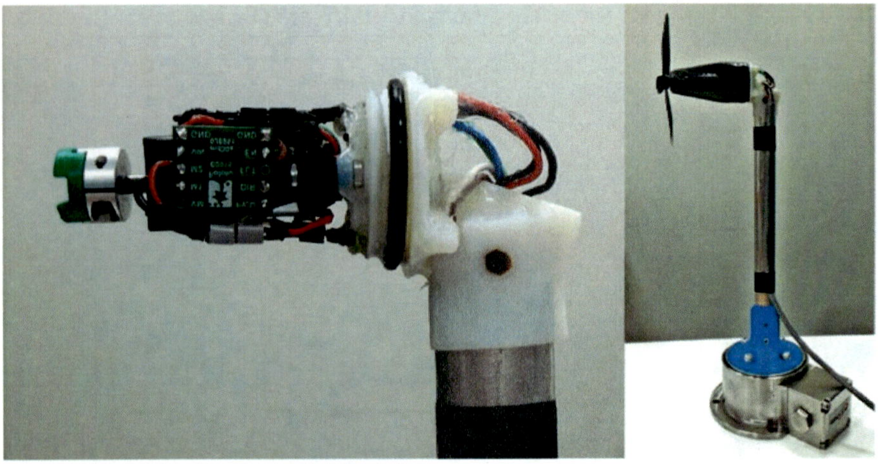

Fig. 10.5 Experimental set up, showing motor and electronics with case removed, and full assembly mounted to a waterproof force sensor

numbers in both media tests were only conducted in water. Since QPROP treats the propellers as inflexible, it was also more important to test underwater to check for any significant effect from blade deformation under the increased hydrodynamic load. In addition, testing in water has advantages over testing in air, as the slower rotation speeds and larger forces in water can be measured more accurately with sensors.

During testing, propellers were driven by a 10 W brushed gearmotor with a 50:1 gearbox (Pololu 50:1 HPCB6V), which could provide the torque necessary to drive the propellers underwater, without drawing damaging current loads, as a large current draw beyond the thermal limits of safe operation when the AquaMAV's propulsion motor underwater.

The encoder mounted to the back of the motor was used to measure rotational speed of the motor shaft, with an accuracy of 600 pulses per revolution of the output shaft. The motor and encoder were contained in a sealed housing, mounted to the force balance via a 30cm streamlined aluminium strut. A propeller drive shaft was passed through the casing using a sealed bearing and connected to the motor output via an Oldham coupling (Fig. 10.5).

The motor power was controlled using pulse-width modulation (PWM), with all control and data acquisition performed in LabView. Each propeller was driven from 10–100% PWM duty cycle, increasing in 10% increments. For every measurement thrust and rpm were recorded for 10 s and averaged.

10.3.1 Results and Analysis

The results of three propeller tests are shown in Fig. 10.6A–C, with thrust curves shown for tests with 2 and 3 blade variants. The computational predictions were found to be a close fit to the experimental data and show that the simulations provide an accurate representation of the actual system. A possible source of the minor discrepancy between the theoretical estimate and actual data is the effect of significantly higher hydrodynamic forces acting on the blade underwater causing slight deformation near the tips. This flexibility was not accounted for in the QPROP simulations. Nevertheless, the experimental results were found to match the simulations closely, indicating that any effect caused by this deformation was not significant. Overall efficiency of the system cannot be concluded from the experiments as only static tests were conducted, meaning that $\eta_P=0$.

This confirms the results from Sect. 10.2 that using an aerial propulsion system directly in off-design conditions will result in highly inefficient propulsion in water, in addition to high current draws for motors not designed to sustain high torque loads. In seeking a compromise between aerial and aquatic performance, the torque requirements to achieve desirable aquatic propulsion would result in a large reduction of thrust in air. As the operating rpm and torque is an equilibrium point based on the motor-propeller combination, a significant improvement in performance can be achieved by using a more flexible system that is capable of altering one of these variables to suit the operating medium.

10.4 Aerial-Aquatic Locomotion

From the propellers investigated, it is clear that using the same motor-propeller combination for multimodal locomotion will be highly inefficient in at least one medium. This problem could be addressed to some extent by variable pitch propellers, but the mechanical complexity makes this challenging to implement for small scale vehicles. A more straightforward alternative is the use of two separate propulsion systems, optimised independently. However, the vehicle would have to carry the weight of one unused system at any point in its mission, and the inactive propulsor may also incur drag penalties.

Because the problem is not that aerial propellers are necessarily inefficient underwater, but that motors are poorly matched, the use of a variable transmission to ensure good matching is sensible. However, this again may be difficult to implement at the small scale, and requires a system for changing gear between media. Rather than employ an actuated gearbox, which incurs a weight and complexity penalty, it is proposed that reversing motor direction is a simple and lightweight means of controlling a two-speed gearbox for an aerial-aquatic robot.

To investigate the efficacy of a transmission system, QPROP has been used to compute an ideal transmission for an aerial propulsion system operating underwater.

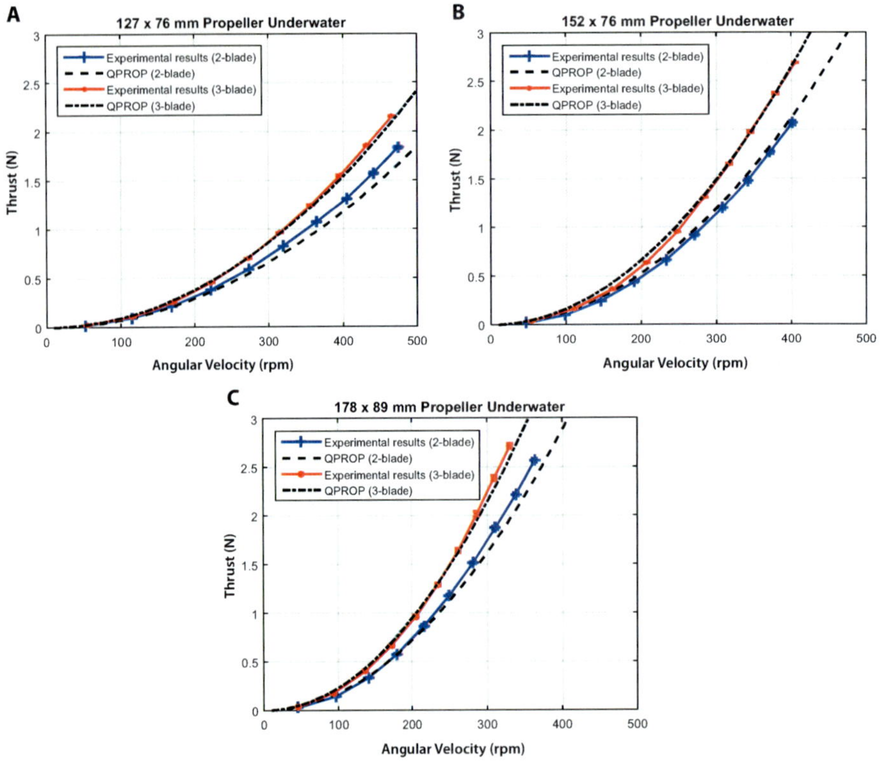

Fig. 10.6 Experimental results for six different aerial propellers used underwater, compared with QPROP predictions, with error bars shown on measured data. Two and three blade variants of three different propeller sizes were tested. Propellers matched well to simulation

In Table 10.1, several propellers are listed and the gear reduction which maximises the best achievable efficiency underwater is computed, where the 40 W brushless motor used in the AquaMAV (Chap. 7) is used as a drive. Across the range of propellers simulated, it can be seen that using the optimal gearing gives an order of magnitude increase in efficiency underwater. Comparing the different propellers shows that large gains can be made by adjusting the torque characteristics of the motor, while varying the diameter and number of blades contributes far less to performance. Under optimal gearing, the underwater efficiency is capable of obtaining a similar range as the aerial efficiency. This means that the system is able to produce significantly more thrust underwater, whilst also drawing a smaller amount of power. Depsite producing less thrust, the power draws required for the ungeared motor exceed the 40 W thermal limit of the motor, while the power required with gearing underwater is of a similar magnitude to that required in air, meaning that underwater locomotion will not place any strain on the electronics of the vehicle.

Table 10.1 Aerial-aquatic performance of different propellers being driven by a 40 W brushless motor (unthrottled): For each propeller,the forward velocity (v) at which the motor-propeller system achieves maximum efficiency is shown. Aquatic performance without variable transmission is compared to the use of a calculated optimal gearing which maximises total efficiency, η_T, increasing thrust (T) while greatly reducing electrical input power, P_i (Note that any P_i >40 W will overheat and damage the motor)

Propeller	Air, direct drive					Underwater, direct drive					Underwater, optimal gearbox					
D × pitch	η_T	η_P	T	V	P_i	η_T	η_P	T	v	P_i	Gearing	η_T	η_P	T	V	P_i
(mm)	(–)	(–)	(N)	(m/s)	(W)	(–)	(–)	(N)	(m/s)	(W)	(–)	(–)	(–)	(N)	(m/s)	(W)
130 × 70 (eflite)	**0.491**	0.680	0.67	15.1	26.5	**0.061**	0.564	3.1	2.40	126.6	13.1:1	**0.497**	0.691	6.8	1.8	24.9
152 × 76 (GWS)	**0.346**	0.500	1.0	11.1	33.2	**0.023**	0.435	3.8	0.80	134.6	10.7:1	**0.352**	0.492	8.6	1.1	26.3
152 × 76 (3-blade)	**0.402**	0.574	1.0	12.4	31.3	**0.031**	0.496	3.6	1.2	133.0	10.9:1	**0.409**	0.568	8.6	1.2	24.2
152 × 102 (CAM)	**0.404**	0.606	1.1	14.8	38.9	**0.024**	0.523	3.2	1.0	135.4	12.1:1	**0.426**	0.594	8.3	1.3	25.3
152 × 102 (APC)	**0.515**	0.739	0.85	19.5	32.2	**0.042**	0.575	2.5	2.2	131.5	11.2:1	**0.528**	0.734	7.2	1.8	24.4
178 × 89 (GWS)	**0.425**	0.619	1.1	13.5	34.9	**0.028**	0.514	3.4	1.1	134.2	11.4:1	**0.437**	0.608	8.90	1.2	25.1
178 × 89 (3-blade)	**0.373**	0.573	1.2	12.8	42.1	**0.020**	0.457	3.1	0.86	135.9	12.6:1	**0.399**	0.558	9.5	1.1	26.2
178 × 102 (APC)	**0.371**	0.592	1.3	13.1	46.3	**0.019**	0.509	3.5	0.74	136.7	13.6:1	**0.413**	0.575	9.5	1.1	25.4
178 × 102 (3-blade)	**0.310**	0.536	1.4	11.8	54.7	**0.013**	0.439	3.2	0.58	137.8	15.3:1	**0.369**	0.515	10.0	0.96	25.9

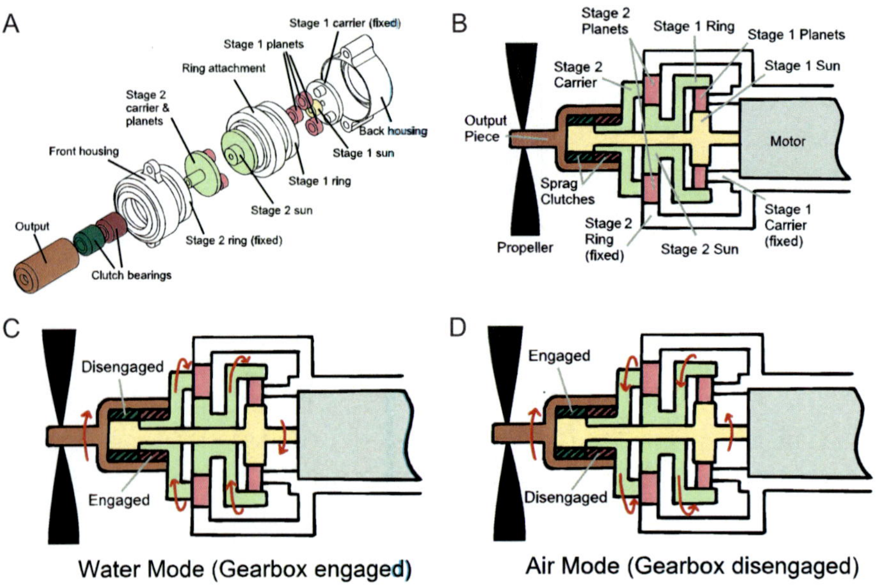

Fig. 10.7 (**A**) Labelled CAD diagram of the developed gearbox. (**B**) Simplified diagram with colors matched to (**A**). (**C–D**) The two operational modes of the gearbox determined by the direction of the motor shaft, with rotation directions marked. *Hand-drawn diagrams by Yu Herng Tan*

10.4.1 Gearbox Mechanism

The simplest way to control the gearbox is to use the drive shaft to automatically engage the gearbox when operating in one direction (water mode) and disengage the gearbox in the other direction (air mode or direct drive), avoiding the complexity of a mechanical gear change, and the need for additional actuators. However, in both modes the output needs to be spinning in the same direction as the propeller is unchanged. In order to achieve this, the gearbox must reverse the direction of the output when engaged, which is done using a planetary gearset in fixed-carrier mode (Fig. 10.7). A second epicyclic stage in fixed-ring mode is added after the first to achieve the full reduction required, while preserving the original direction change. Both the output and input driveshaft to the gearbox are connected to the propeller using separate sprag clutches, which permit rotation in only one direction. Because the output of the gearbox spins at a lower speed than the driveshaft, the two move relative to each other and are selectively decoupled from the propeller by the clutches. This means that while the gearbox can be driven by the motor in both directions, the propeller spins with the input when the motor rotates in one direction and with the gearbox output when it rotates in the other.

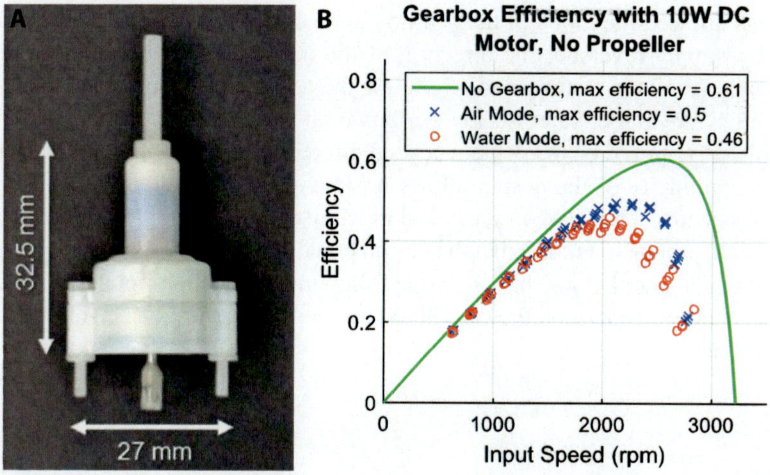

Fig. 10.8 (**A**) Manufactured prototype of the gearbox. (**B**) Efficiency data for fabricated gearbox being powered by a 10 W DC motor, compared to the motor characteristic

10.4.2 Prototype

A gearbox prototype was produced which weighs 12 g, is 27 mm wide, and 32.5 mm long (Fig. 10.8A). Miniature steel gears were used for the transmission, while the gear housing and output connection were 3D printed using a Connex Objet350 printer. Using the 13 × 70 flight propeller, our analysis (Sect. 10.4, Table 10.1) gives an optimal gearing of 13.1:1, and as a result our gearbox is designed to obtain a gear reduction of 12.7:1, with a slight difference due to the availability of appropriately sized gears.

This gearbox was run across a range of speeds (using PWM) by a 10 W DC motor (Pololu 10:1 HPCB6V), and by measuring the difference between the measured speeds of the motor with and without the gearbox attached, it was possible to calculate the resistance torque the gearbox creates at a given input speed (using Eq. 7.4), and thereby the loss of efficiency (Fig. 10.8B). The tests showed that the maximum achievable efficiency of the motor was reduced by 18% when operating with the gearbox disengaged (air mode) and by 24% with the gearbox engaged (water mode). The friction losses are similar as gear train rotates in both modes, but in air mode, the input shaft rotates in the opposite direction to the stage 2 carrier (Fig. 10.7D) so the increased relative motion results in higher friction. The friction in the drive is slightly higher than anticipated, due mainly to the poor alignment achieved using 3D-printed plastic to house the transmission, which can be readily improved in a future prototype.

A prototype system for reconciling the differences between aerial and aquatic propulsion was demonstrated through the integration of a specially designed gearbox. The novel use of a two stage planetary system with paired one-way clutch bearings allowed the mode of the gearbox to be controlled with no additional actuators by simply reversing the direction of the motor. The final prototype showed promising results in testing, and will allow the aerial propulsion system to operate more efficiently underwater. As the gearbox is an independent add-on, no additional controls are required on the vehicle beyond directional control of the existing motor, and the simplicity of the design allows it to be easily integrated into other aerial-aquatic systems. The gearbox presented maximises the underwater efficiency of the AquaMAV propulsion motor, but the design process has otherwise been independent of the vehicle detailed in Chap. 7. In the following chapter the impact of the aerial-aquatic gearbox system on the overall locomotion energetics of the AquaMAV will be examined.

Chapter 11
Sailing and Flying with a Multimodal Robot

11.1 The Challenges of Surface Locomotion

The field of aerial-aquatic robotics promises tremendous benefits in data collection as well as unmatched flexibility and remote access. However, the majority of existing aerial-aquatic robots are unable to perform scientific tasks at significant depth, limited by the weight penalty that any pressure resistant container would add. In addition, sealing of an actuated robot is difficult, again adding significant weight to small systems. Wireless communication is a major challenge for underwater robots and certainly poses great constraints to operation at distance. Lastly, underwater propulsion is often highly inefficient due to geometries optimised for flight [109]. Indeed, most aerial-aquatic vehicles either have severely limited water range and operation, stay in very shallow waters or function only in de-ionised water. Too often, the benefit of underwater locomotion is overshadowed by the weight, and complexity increases that are required for reliable operation. This negatively impacts flight performance.

Shallow water monitoring is also achievable with vehicles that move at the water surface. Those robots save on many of the above-mentioned difficulties and instead provide a broader range of features, such as longer, more efficient operation, higher sensor payload, live communication, GPS positioning and faster deployment time due to a more modular design and more.

Fixed or underwater only solutions also exist, such as autonomous water data acquisition systems that are being used in oceanography nowadays. The most prominent platforms are the Argo Floats [185], the Seagliders [186] and the Saildrone [187, 188]. All these platforms have long endurance but are heavy, costly ($20,000 and 30 kg per Argo float to $100,000 and 50 kg per Seaglider) and cannot be moved easily (or at all).

Quadcopters have become a widespread tool in marine research but their mission profiles are limited because of short range and flight time. In order to overcome those limitations, we propose here a new, efficient fixed wing configuration that sails through the waters in a very low power mode and will therefore be able to recharge using solar panels, meaning the technology is not only amphibious but also has the

R. Zufferey et al., *Between Sea and Sky: Aerial Aquatic Locomotion in Miniature Robots*, Biosystems & Biorobotics 29, https://doi.org/10.1007/978-3-030-89575-4_11

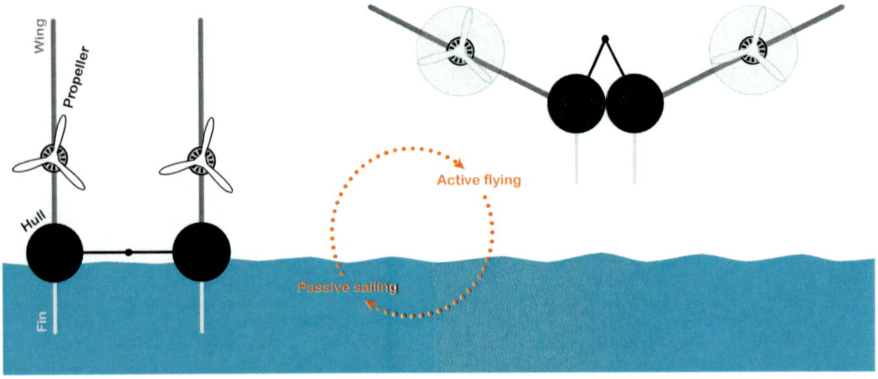

Fig. 11.1 Sailing-flying locomotion concept with dual purpose wings

potential to be energy autonomous. The possibility of using the robot in long range sailing missions, opens even more possibilities, and the range limit imposed by the radio controls can be safely exceeded.

Sailing offers a way to navigate on the surface with minimal energy expenditure. Thrust is entirely generated from wind power, leaving control and sail actuation the only power draw onboard. For example, the Americas cup boats achieve 2.2 times the wind speed and reach speeds of over 80 km h^{-1}, based on catamaran hull and rigid sails [189]. The potential of wind powered robots has already captured some attention, since they are an affordable, long duration and sustainable form of locomotion [190, 191]. One of the main challenges of using sails for an autonomous drone is developing the control mechanism to ensure correct navigation, however, some successful implementations have been shown [192] using simple sensors.

This chapter presents a new class of aerial-aquatic robots (Fig. 11.1) that can perform direct aquatic measurements at the surface whilst offering the flexibility and range of flying. The flying-sailing robots exhibit multiple take-offs, transition to flight and passive locomotion on the water surface. They should leverage environmental vectors, i.e. wind to propel themselves through the water and solar energy to recharge its batteries for the next flight. In this way, long-duration missions become possible with only minimal energy usage, while at the same time the benefits of flight can be fully exploited.

11.1.1 Cycle

A key requirement in the development of a lightweight hybrid aircraft is the reduction of complexity. This is addressed by employing the same control surfaces for flying as for sailing. Despite the three order of magnitude difference in density between air and water, the same surfaces, actuators and electronic configuration are used for both air and water mobility.

Table 11.1 Aerodynamic surface use in hybrid design

	Servos	Flying	Sailing
Central wing	0	Lift	Structural
Sails	2	Lift Roll control Flaps control	Propulsion
Elevator	1	Pitch control	
Rudder	1	Roll control	Heading control

The proposed robot concept features a three-part wing, i.e. a *central wing* terminated on its extremities by two variable-pitch wings. The latter two surfaces pivot upwards during sailing and will be referred to as *sails* for clarity, even in flight mode. The central wing possesses increased efficiency compared to the sails due to its non-symmetric aerofoil and boxed section that reduces lift induced drag occurring at wing tips. The tail is elevated above the water when sailing, however the rudder reach into the water and serves as heading control during sailing. The duality of aerodynamic control surfaces is summarised in Table 11.1.

The SailMAV[1] is designed to execute a repeated sail-fly-sail cycle. It *sails* on the surface of the water, propelled by the wind, whilst fulfilling its scientific mission e.g. water monitoring and recharging its small battery. During water locomotion, an autonomous sailing controller ensures that the robot can move to specific waypoints. When a change of body of water is required, a large distance needs to be covered or an obstacle obstructs its way, the vehicle switches to *flight* mode. The take-off and flying phases use a standard fixed-wing flight controller. The vehicle can later land on water again for its next sensing mission.

11.1.2 Hybrid Sizing

The design of a hybrid system depends on many factors that can have positive impact in one operation mode while having negative influence in the other. The driving factor for this design was to retain the lightest structure possible while employing off-the-shelf components. Most of the original design stems from lessons learned from the AquaMAV robot which weighs 150 g. The addition of more servo-actuation, control and data collection has increased the total mass budget to 500 g. The resulting wingspan is calculated from that, considering a high-lift central wing and low-lift symmetric sails. A cruise speed of a maximum of 10 m/s is tolerated as an arbitrary factor. Good sailing performance is possible with large sails. However, this comes at

[1] The acronym MAV was originally suggested for flying vehicles below 15 cm by DARPA. However, the term has by now been used extensively in research to describe small-sized lightweight aircrafts that exceed this dimension but are still significantly smaller than planes.

the expense of reduced flight lift (for the same total wingspan) and reduces stability as the hulls are closer together. In this presented concept, the sailing performance is not maximised, rather preferring a more capable flying vehicle. As such, the sails only represents 50% of total wingspan. This gives a robust design which is stable and simple to manufacture (no wing longer than 50 cm) with a larger high-lift central wing.

11.1.3 Propeller-Powered Takeoff

The aquatic escape is the most challenging part of the whole mission. There are two possibilities of propeller-powered transition from surface to flight (Table 11.2).

Seaplane-like take-off is composed of three phases, as described in the seaplane literature [193] and shown in Fig. 11.2. First, the propellers are turned on and the robot is in a low-speed regime, i.e. the hulls are in the water and the plane behaves as a displacement vessel. As the speed increases, the robot moves to a planing range (phase 2), passing the point of highest water resistance called the hump. When planing, it touches the water only at the afterbody section of the boat. In this topology, planing is assisted by the central wing, which increases the lift substantially through *ground effect*, providing up to 28% additional lift, see Sect. 11.5. Thirdly, full flight is achieved after take-off and the aircraft behaves as a standard fixed-wing vehicle. The wings provide increasing lift with higher speeds. As such, propelled acceleration at the water surface will enable take-off given sufficient thrust. Water drag needs to be overcome, which increases until *hydroplaning* velocity. At that speed, the aircraft is gliding on the surface and drag is reduced.

Vertical Take-Off and Landing is another method for leaving the water. The rotors are tiltable and oriented so that the thrust vector is vertical. At rest, the motors are turned on to full thrust until the aircraft vertically takes-off, much in the same way commercial quadcopters operate. The wings do not generate any lift nor do they offer control. Therefore, attitude control of the aircraft can only happen through thrust control. This method might be interesting for future developments of the SailMAV concept, due to the necessary actuators (wing pitch control) already in place.

Table 11.2 Lift-off versus VTOL water-to-air transition (WAT)

	Lift-off	Water VTOL
Principle	Aerodynamic surfaces create lift	Propellers create sufficient lift
Cost of method	Take-off distance required	Thrust of motors > Weight, large battery
Limitation	Surface only, sea state may hinder take-off	Difficult control, development time

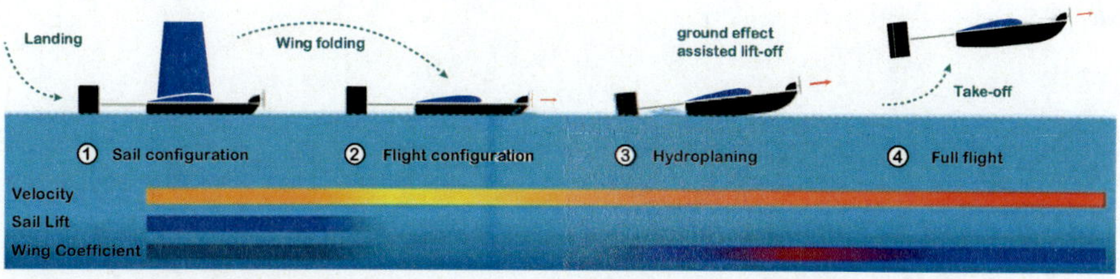

Fig. 11.2 Sailing-flight cycle overview. The robot lands on water where it transitions to a *sailing* configuration and executes its mission. Take-off is achieved via propelled acceleration, transitioning from displacement vessel regime, to hydroplaning and lastly to being fully airborne

11.2 Harnessing the Wind

Autonomous motion at the water surface is today an existing technology that permits ultra-long missions. Saildrone [190], for instance, is currently autonomously roaming the ocean for marine investigation, for applications such as animal tracking and environmental monitoring. At 750 kg, however, this robot, like most AUVs, is constrained to large bodies of water and long-duration missions. At this scale it would be ill-suited for small bodies of water, shallow waters or shorter missions where fast results and deployment are critical.

Relevant physics at play in nature are briefly presented here, as a proper understanding thereof is required for adequate sizing.

The main forces at play in sailing are aerodynamic and hydrodynamic forces, both of which depend on the velocity of the boat. The sailing surface—whether a sail or a rigid wing—is subjected to the apparent wind velocity V_{AW} (Fig. 11.3A). The apparent wind is the true wind velocity V_{TW} translated into the frame of reference of the craft which moves at a given speed V_B. The apparent wind is defined as $\vec{V}_{AW} = \vec{V}_{TW} - \vec{V}_B$. For low boat velocities, solely considering true wind will give acceptable sailing performance.

The sail generates a total aerodynamic force F_T. It can be decomposed as a lift force F_L perpendicular to the apparent wind and a drag force F_D orthogonal to the lift. Both forces depend on wind speed V_{AW}, density of the air ρ, area of aerofoil A_{sail} and coefficients C_x, determined through experiments or simulation.

$$F_i = \frac{1}{2}\rho A_{sail} C_i V_{AW}^2 \qquad i = L, D \tag{11.1}$$

The drag force comprises skin friction drag and lift-induced drag. F_T can also be projected into the boat frame as drive force F_{DRIVE} and lateral force F_{LAT}. The drive force moves the craft forward while the lateral force will tend to shift it sideways.

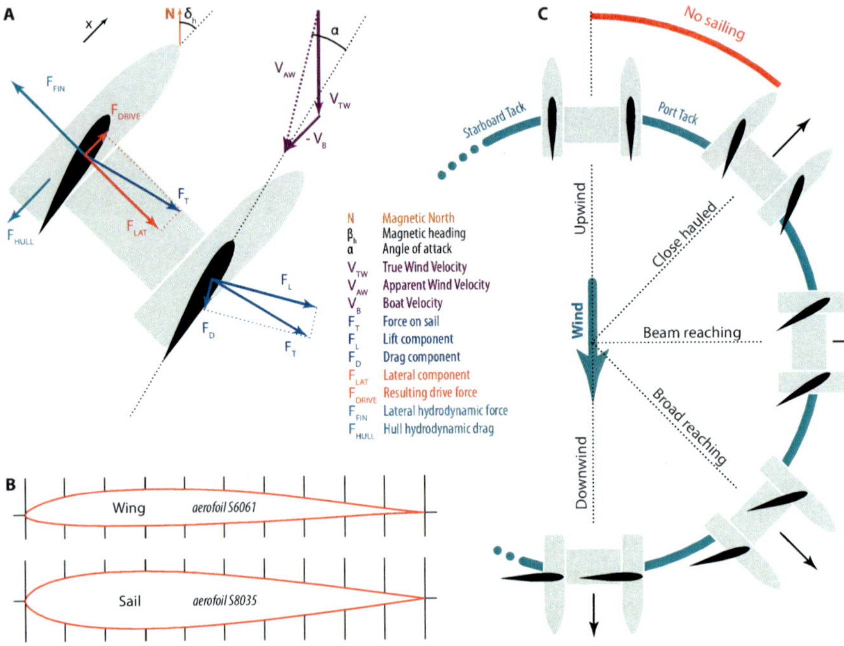

Fig. 11.3 **A** Forces, velocities and angles relevant in sailing. Forces are distributed between the left and right hull for clarity, they are identical in reality. **B** wing profile choose for the asymmetric wing and symmetric sail. **C** Diagram of the sail position in function of wind/boat orientation

F_{LAT} needs to be compensated by hydrodynamic forces on the fins, hulls and keel. This is effectively a high water drag force due to the large surfaces involved (lateral consideration). Finally, all surfaces under the water line will have a hydrodynamic drag labeled F_{HULL}. This last force increases with boat speed until hydroplaning occurs.

The sailing craft's motion (Eq. 11.2) can be solved knowing the coefficients C_L, C_D and C_H, the hull drag coefficient. The sail surface will determine the steady-state velocity of the boat, for a given wind speed and angle.

$$m\frac{dv_x}{dt} = F_L sin(\alpha) - F_D cos(\alpha) - F_{HULL} \tag{11.2}$$

$$\text{where} \quad F_{HULL} = \frac{1}{2}\rho_W A_W C_H V_B^2 \tag{11.3}$$

The aerofoil S8035 (Fig. 11.3B) selection is a trade-off of various constraints. A thick profile permits integration of servo actuation directly within the thickness and minimizes the manufacturing inaccuracies that occur at the leading edge (i.e. it is rounder). A symmetrical profile enables equivalent sailing in both port tack (wind on the left) and starboard tack (wind on the right). Lastly high lift at low Reynolds number is the maximising factor for best performance. The central wing of the robot features an S6061 airfoil which has better characteristics, because it is optimized for flight only and does not integrate any additional components within.

A sailing watercraft can travel in any direction between close hauled and downwind with respect to the wind (Fig. 11.3C). Immediately, it is clear that in order to sail, a human or an autopilot controller requires knowledge of the wind direction, from experience or from a sensor. A boat cannot move upwind, where $F_{DRIVE} < 0$, indicated in the wind diagram. This limit angle varies with speed and hull/sail design and is usually situated between 30° and 60°. Figure 11.3C shows a sailing catamaran in port tack orientation, behavior is mirrored in starboard tack.

An interesting technology which could be available to wind-power autonomous robots is kiting. Kite usage has recently seen an explosive growth in the sports industry as well as being a power source for boats [5] and in power generating plants. For example, kites have been introduced in large ships to generate electricity as well as to propel the boat forward using the ladder-mill concept [11]. This idea could serve as an example for both recharging and propelling an Aerial Aquatic Vehicle drone. However, the control of kites is challenging. While successful attempts at making a controller for a kite have been made [12], miniaturising the system and ensuring that the kite remains in the air at all times, even at low wind speeds may prove to be too complex. Powered flight with a kite-wing also comes with its set of difficulties, although powered paragliding proves the validity of the concept.

Altogether, the use of rigid wing surfaces simplifies control, design and manufacturing without reduction of efficiency. Indeed, kites are a more powerful energy source because they leverage stronger, more stable winds at higher altitudes than equally sized sails. This, however, is irrelevant for this application where longevity rather than propulsive power is the design driver. Additionally, at small scales, lower wind speeds are desirable.

11.3 Design

An aerial-aquatic robot, even in the case of surface contact, poses substantial opposing constraints. Its weight needs to be minimised for take-off yet the aquatic environment requires adequate waterproofing as well as bulky buoyancy elements that negatively affect flight. Furthermore, a research robot should be highly modular for adjustments and continuous upgrades, which also increases its weight. The next chapter discusses how to balance these elements effectively to obtain a 520 g field-ready autonomous hybrid vehicle.

The proposed vehicle is built around a *catamaran* structure with a three-segment wing. The central part (*wing*) of the wing is fixed to the hulls which serves a double function as aerodynamic surface and structural link. The two outer parts of the wing (*sail*) can change their orientation to act as rigid sails when the aircraft is on the surface of the water as shown in Fig. 11.4.

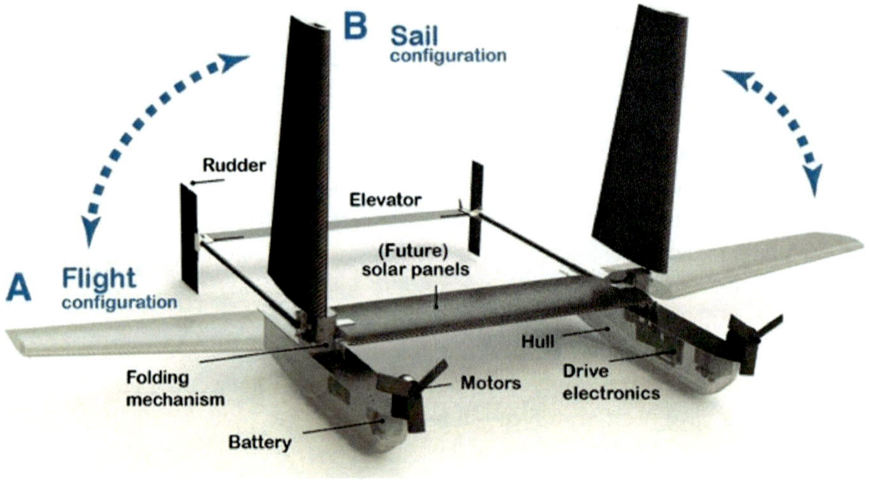

Fig. 11.4 Schematic clarifying the two modes of operation of the robot. (**A**) The sail in flying configuration, showing a 6° angle to the horizontal. (**B**) The sail is moved to an upright position, for the sailing configuration

The SailMAV is composed of a carbon fibre, foam and shells structure. As the robot is a research platform, each of its component should be replaceable and upgradable, without compromising waterproofing.

The two hulls are linked by a rigid closed-cell wing, with a high-performance foam core and wrapped with one layer of 100 μm carbon fibre fabric via a wet-layup process. This technology provides outstanding strength and resilience to crushing with a short manufacturing cycle of approximately one day. Two sails are built using the same methods, but additionally feature built-in servo actuation to change their Angle Of Attack. At the end of each hull two carbon-fibre tubes are affixed, which compose the tail and carry the drive wires for the control surfaces.

11.3.1 Morphology Change Actuation

One of the key challenges of this new aircraft geometry is to develop and integrate a mechanism allowing the morphology of the sails to change. A crucial requirement to this actuator is non-reversibility, as both modes will be enabled for prolonged periods. Integrating a small, light-weight, compact, reliable, high torque, waterproof-able device presents formidable challenges, further described in Sect. 13.

As shown in Fig. 11.5, by rotating around the longitudinal body axis, the sails can assume two different orientations (to horizontal): 90° for sailing, i.e. exactly vertical, and 10° for flight. This dihedral angle of 10° improves roll stability in flight configuration and can be changed at will by replacing the cam. Moreover, in both

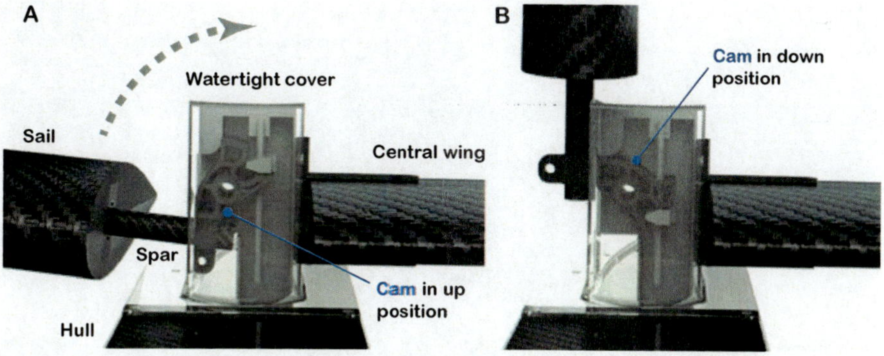

Fig. 11.5 **A** The sail in flying configuration, showing a 6° angle to the horizontal. **B** The sail is moved to the upright position, for sailing. This mechanism is double locking, the central cam is irreversible at both ends of the lead screw. *CAD render by C. Raposo*

modes, each sail can rotate around itself, to control the plane in roll (flight) and to orient the sails (sailing), detailed in Sect. 11.5.2.

11.4 Avionics

The SailMAV robotics platform has a novel structure and hybrid functionality which requires a unique avionic design to achieve autonomous operation.

Overview The robot is controlled by a Pixhawk[2] flight microcontroller running the *PX4* firmware. The inclusion of a dedicated autopilot controller significantly expands the autonomy of aerial-aquatic robots, limited in the prototypes presented in previous chapters. Featuring an ARM M4 running at 180 MHz, the autopilot board outputs the signals to the 8 actuators and reads all onboard sensors. It is connected in the left hull to a power and signal PCB, custom-designed for the robot. This board interfaces with the *central wing bus* (CWB), a 15-line link to the right hull. In addition, the folding actuator as well as the tail and sail servo are also connected to the board. The CWB is made from marine grade ultra-thin silicone-coated wire that can resist harsh conditions (see Fig. 11.6).

Two batteries resting in each hull are connected in parallel for mass symmetry, feeding 16 V voltage to the ESCs and the voltage regulator which powers both the controller and the actuators independently. Both the 16 V and regulated 5 V lines are distributed on both hulls via the CWB.

Three sensors are located onboard, sharing data for control. A GPS receiver with barometer ensures stable flight in outdoor tests and offers room for expansion towards waypoint-based control. Wind velocity is obtained from a pitot tube air speed sen-

[2] Pixracer R14 from mRobotics.

Fig. 11.6 Upside-down view of the composite wing with the Central Wing Bus passing through the centre of the foam core. The end connectors are only added after wing curing

Fig. 11.7 PCB Designs of the left and right side electronics boards, carrying the flight controller and the data acquisition system

sor. Lastly, a new wind angle sensor is necessary for autonomous sailing. To the best of our knowledge, no such lightweight wind angle sensor exists. A lightweight, integrated sensor is developed for this application which interfaces with the flight controller, as detailed in Sect. 11.4.1.

Main board SailMAV features two identical hulls which contain all electrical system except for the sail actuation. For flight stability purposes, the robot's weight should be equalised on both sides. As batteries overshadow the mass of all other internal components, they are doubled and equally placed in the front of each hull.

The right hull features a PCB that hosts *flight* components shown in Fig. 11.7 right. This includes the controller with its array of on-board sensors, communication links and antennas, external sensors required for hybrid locomotion such as GPS, wind angle and differential pressure. A 10W regulator produced the 5 V bus that supplies all sensors and actuators.

The left hull carries the peripherals needed for *data acquisition* as well as additional servos. Modularity is a key driver in this design and as such, multiple data

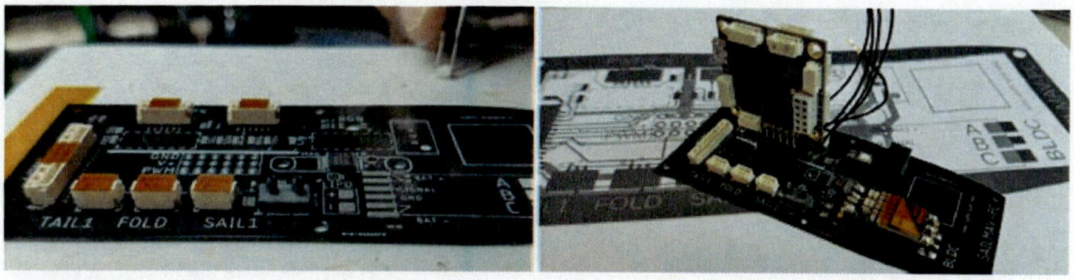

Fig. 11.8 (left) Bare PCB board getting populated with components. The reflow paste is visible in grey as well as the differential pressure sensor in the front. (right) The flight controller is connected directly into the PCB-FC, resting on a print of the board layout

sensors can be connected over a shared I2C bus. An on-board atmospheric sensor rests on the inside and is linked to the outside air via a silicone tube through the PCB. The acquisition of data is handled by a separate ARM M0 microcontroller (Adafruit SAMD Mini). Data is locally saved onto a micro-SD card slot, situated next to the microcontroller. Both are visible in the top part of Fig. 11.7.

The robot possesses a unified connector system, the JST-GH series, visible in Fig. 11.8. This series is excellent for lightweight robotics with its 1 mm pitch, adequate power transmission with 1A capability, and latches which guarantees a trustworthy connection in flight. In addition, the external breakout boards (e.g. controllers) are all plug-and-play in case replacement or upgrade is required.

Propulsion Recent advances in the efficiency of switching solid-state electronics have hugely expanded the use of brushless DC motors. Now present in most drone technology, small and large aircrafts, electric bicycles and cars, they boast high reliability, inherent water resistance, low rotor inertia and high speeds. A brushless motor consists of a static set of coils driven by 3 high current lines, i.e. the *stator*. The *rotor* of small motors is in general a permanent magnet that will follow the magnetic field imposed by the coils. This type of motor requires electronic commutation of the three phases. Due to high switching losses of transistors brushless motors have only become available recently. Nowadays, the commuting efficiency of MOSFETs is high.

Figure 11.9 shows the connectivity of brushless motor with the power source. A 25 A Electronic Speed Controller (ESC) with a large current margin is selected, allowing freedom of motor selection and battery voltage. The high current carrying cables from the battery to the ESC and ESC to the motor are minimised in length by design.

The choice of the motor is highly related to the choice of the propeller and the batteries. Indeed, a motor is usually optimised for a specific size of propeller and the current drawn by the motors matters for the choice of the batteries. All these three components should therefore be tested together. Experimental specification validation (1520 series RC benchmark static thrust stand test) often shows differences with the published specification, especially with low-cost brands. Based on experimental results, a RS1306 v2 2700 kv motor was selected with 200 g of measured static thrust,

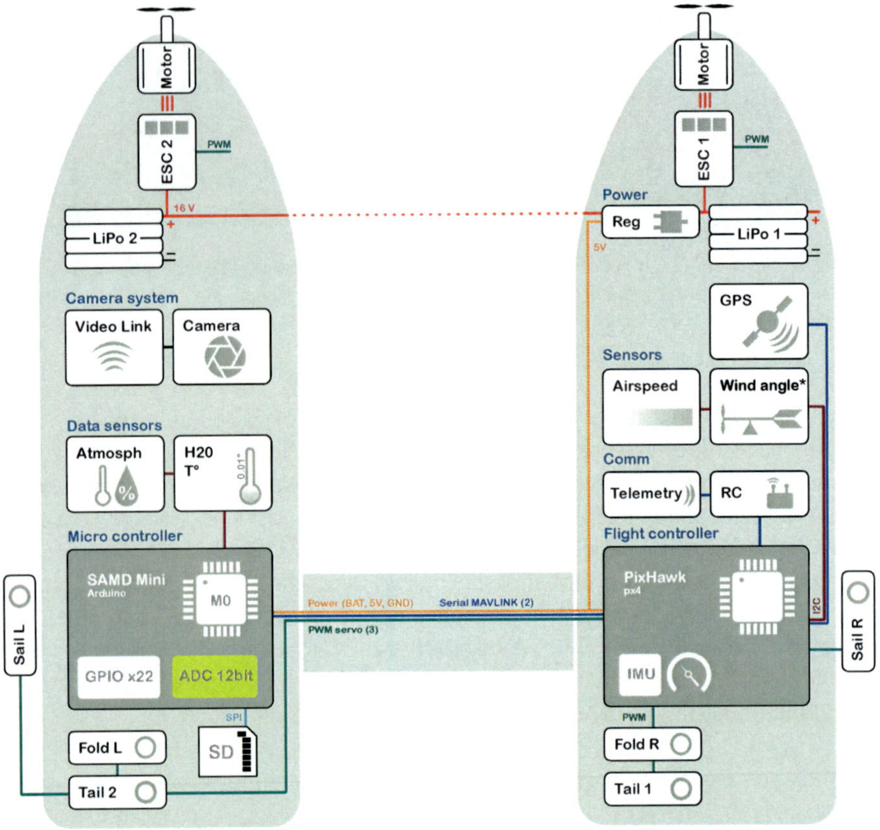

Fig. 11.9 Simplified electronic peripheral devices layout and their connection. A central PixHawk, located in the left hull communicates with 3 sensors for controllability, 2 communication links and 8 actuators in total, spread between the two hulls. The device is powered by four cell 300 mAh lithium polymer batteries, one in each hull. The wind angle sensor (*) is in development

operating at 4 cell voltage and a 2.1 A current draw at cruise speed. This ultimately determines the flight range and should be minimised. The overall drag coefficient of the aircraft is calculated in CFD to be $C_D = 0.08739$[3] which gives the required thrust at cruise to be of 0.93 N. Some margin is required as the thrust in flight is lower than the static thrust. This is due to the thrust generated by the propellers decreasing with speed. This selection gives a maximum of 400 g of tested thrust, more than theoretically required for take-off.

[3] Simulation results obtained from A. Ortega.

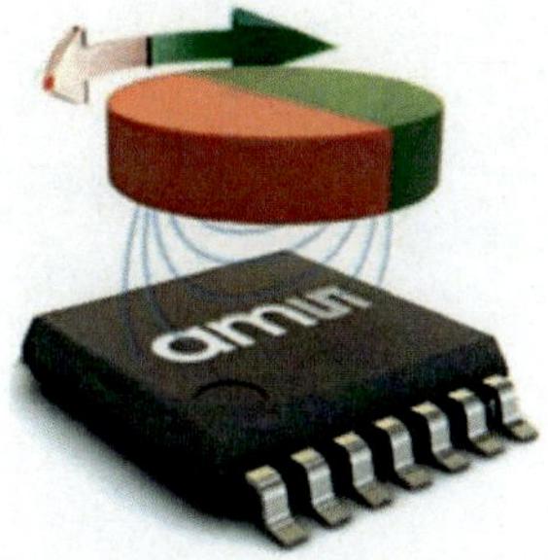

(a) Rendered view of the field generated by a permanent magnet read by the AS5048b sensor. Source: AMS.com

(b) Sensirion SDP3X Sensor. Source : sensirion sensor company

Fig. 11.10 External sensors required for fixed-wing flight and sailing

11.4.1 Wind Sensing

The wind angle is obtained by reading the angle of a small 6 mm magnet that spins axially over/under a digital magnetic encoder (Fig. 11.10a). The encoder chosen is a AS5048B sensor. This sensor comes in a small surface mount device format and is soldered in the centre of the PCB, on the inside. The chip features four hall effect sensors and logic that converts the measured magnetic fields into a clean angle value corresponding to the physical angle of the diametrically polarised magnet. The magnet is held on the reverse side of the PCB—i.e. outside—via a small bearing and a 3D printed ABS casing. It is linked to a mylar arrow that floats downwind.

Fixed wing vehicles rely on the lift generated by their wings to maintain altitude and control. The lift is proportional to the square of the airspeed. In order to avoid stall situations, the system should know the current airspeed of the vehicle. On-board positioning. i.e. GPS and inertial measurements only yield true speed to ground, which does not account for any wind. The smallest yet suitable wind sensor available on the market is the SDP3X (Fig. 11.10b). It is directly surface mounted (SMD) and has a minimal footprint and weight penalty. Moreover, it is already supported by the PX4 framework, reducing development time. The sensor measures *differential* pressure between two ports. These connect with a pitot tube via 2.5 mm silicone airlines. The pitot tube placement is important, it should see an undisturbed air flow. For that purpose it rests under the central wing.

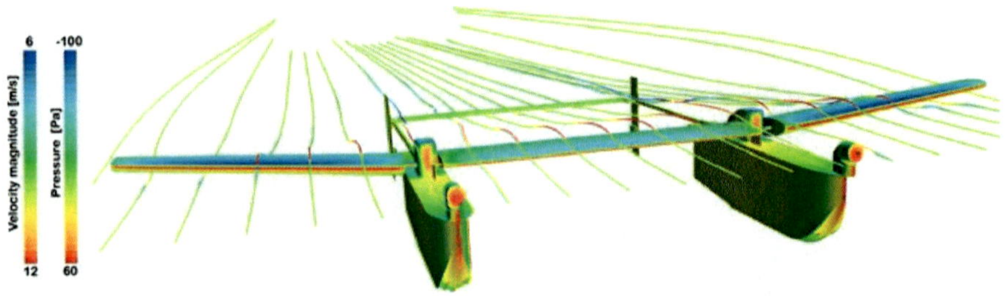

Fig. 11.11 Static pressure contours on the surface of SailMAV, with streamlines colored by velocity magnitude for a free-stream velocity of 10 m/s and central wing angle of attack of 5.75°. Figure generated by A. Ortega

11.5 Aerodynamics Surfaces

11.5.1 Wings

An effective aerodynamic design is needs to provide sufficient lift for take-off and cruise and to minimise drag, in order to achieve a long range and endurance, while maintaining adequate flight stability and manoeuvrability (Fig. 11.11).

The stall velocity is given by (11.4), occurring at the maximum lift coefficient $C_{L,max}$ of the wing for a given wing area and robot mass M.

$$V_{stall} = \sqrt{\frac{2gM}{C_{L,max}A_{wing}}} \tag{11.4}$$

The central wing is rectangular with a S6061 cambered airfoil, selected for to its high lift-to-drag ratio of 51.2 and good stall characteristics at the flight Reynolds number range of 80,000 to 120,000, while remaining possible to fabricate using the foam core manufacturing process (minimum thickness limit). It is fixed at an angle of 5.75° to the hulls, which corresponds to the maximum lift-to-drag angle of attack for the chosen airfoil. All parameters required for wing construction are presented in Table 11.3.

11.5.2 Sails

The lateral wings are used in both flight regimes which makes them complex to actuate. In order to commute from one position to the other, a transmission needs to pivot the sails from the *vertical* (sail) position to the *horizontal* (wing). During sailing, the lateral wings must be actuated to change their orientation depending on

Table 11.3 Characterisation of sails and wings

		Wing	Sail
Aerofoil		S6061	S8035
Chord	c	0.150 m	0.120 m
Span	b	0.35 m	0.25 m
Area	S_{LW}	0.0525 m	0.324 m
Aspect Ratio	AR	2.33	2.2041
Taper ratio	λ	0	0.6
Cruise AOA	α_a	5.75°	7.75°
Stall angle	α_S	10.75°	11.75°
Max lift to drag	$\frac{Cl}{Cd}$	51.2	38.7

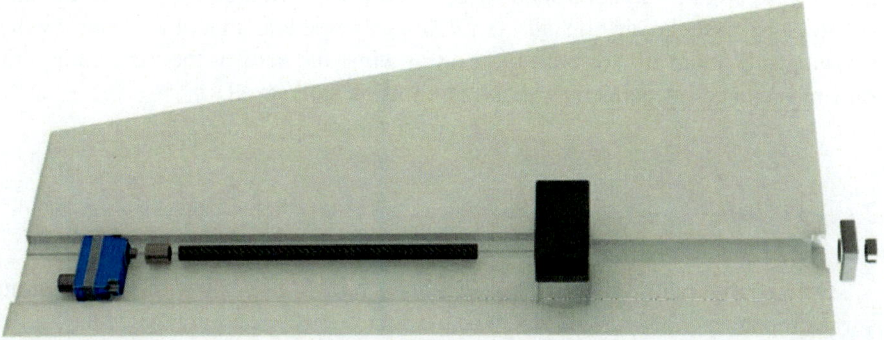

Fig. 11.12 Exploded view of the construction of the sail in horizontal flight configuration. CAD render by C. Raposo

Table 11.4 Sail AOA actuation

	Weight	Torque	Angle	Thickness	Resolution	Voltage
BMS-A10V	7.4 g	2kg cm	± 60°	8 mm	12 bit	5 V (8.4 V)

the wind direction, whereas during flying wings should include an aileron mechanism to control the roll of the plane.

A fitting solution for both flight regimes is to use a servomotor to actuate the sail around the sail spar (Fig. 11.12). As such, the wing rotates and can be used as *wingeron*.[4] A thin servomotor with low weight and high torque is required. A parametric study elected the above-mentioned servomotor which features outstanding capabilities (Table 11.4).

The actuation of automated sails of model or full-sized boats can be composed of two separated airfoils linked by a mechanism which changes the *camber* of the sail.

[4] Combination of wing and aileron.

This produces a similar amount of lift as an asymmetric airfoils [194, 195]. Another method, rather seen in smaller crafts is to actuate the angle of a trailing tail, directly attached to a freely rotating wingsail [190, 196].

However, these solutions were not retained because their actuation system is more complex at the small scale. In addition, the small size of this sailing aircraft reduces the forces at play and can be handled by a servo directly actuating the Angle Of Attack. Perhaps most importantly, this direct drive configuration permits the broadest flexibility, any sailing angle and room for future expansion. Sail-mounted rotors could even permit VTOL capability.

The wing-sails have a symmetrical S8035 airfoil (Fig. 11.3B) due to its efficiency and high stall angle at low Reynolds numbers as well as its large thickness for servo actuation. They have a taper ratio of 0.6 which reduces lift induced drag, structural loads at the root as well as loads on the folding mechanism. The angle of attack of the sails is variable and controlled by the servo, set by default to 7.75°. That value is chosen because the airfoil stalls at a 2° higher angle than that of the central wing, allowing maximum lift for both the central wing and sails at the same time, thus maximising take-off performance and minimising stall speed.

11.5.3 Tail

A stable configuration aircraft conventionally relies on a tail with two actuating surfaces, an *elevator* and a *rudder*. The elevator controls the pitch while the rudder controls the yaw. The tail was designed to work effectively in both air and water. The H-tail configuration, i.e. with the elevator between the two booms adds to the structural rigidity. The elevator is only driven by a single actuator. The tail consists of a 0.5 mm carbon fibre plate with half-round 3 mm carbon fibre stiffeners for light weight and ease of manufacture. The fins and rudder extend downward from the tailplane, as well as upward, so that they are partially submerged, ensuring sufficient control during sailing and take-off. The main disadvantage of this tail design is the coupling of the roll and yaw attitudes of the aircraft. Indeed, when moving the sails, the aircraft will roll but also subjected to an adverse yaw moment. The flight controller however will counteract this effect with the corresponding rudder compensation.

The tail is actuated by two servos, located in each hull. One moves the elevator for pitch control, while the second actuates the rudder. The rudder controls the robot's yaw orientation both in flight and in sailing.

11.5.4 Numerical Simulation Results

The following numerical results were generated with StarCCM by A. Ortega and A. Farinha and serve as a basis for the geometry design of the vehicle. Wind tunnel tests would be relevant in order to supplement and validate these results. Figure 11.13A

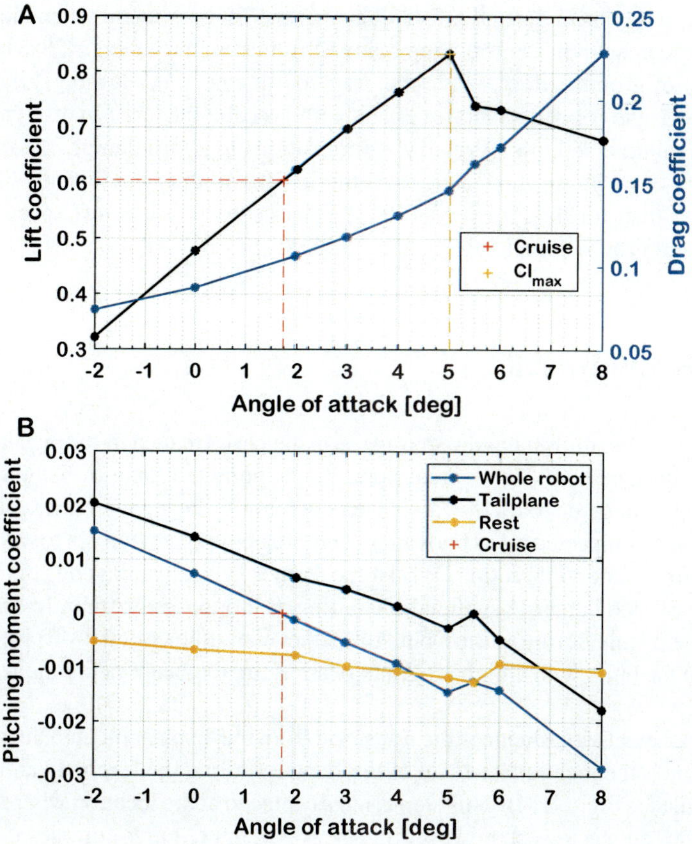

Fig. 11.13 **A** Lift (black) and drag (blue) coefficient variation with angle of attack of the plane, showing the cruise condition at 1.8° and stall at 5°. **B** Pitching moment coefficient about the center of gravity for the whole robot, tailplane, and robot minus tailplane. The plot shows that the SailMAV is trimmed at the cruise condition and that it is longitudinally stable, presenting a negative slope for pitching moment, except immediately after stall. Note that the wings and sails are mounted at an angle with respect to the plane which is taken into account here

shows the lift coefficient (C_l) and drag coefficient (C_d) as a function of the angle of the SailMAV's hull and tail assembly. The lift increases up to an angle of attack of 5°, when the robot stalls. At this point, the C_l is 0.833, which corresponds to a stall speed of $9.3\,\text{ms}^{-1}$. The lift coefficient at the cruise condition is shown to be 0.605, which yields a cruise speed of $10.8\,\text{ms}^{-1}$. Using the C_d of 0.102 for the cruise condition, the drag, and therefore required thrust, for cruise are calculated to be 0.86 N.

Figure 11.13B shows how the pitching moment coefficient about the centre of mass changes with the angle of attack. The SailMAV is shown to be trimmed, requiring no elevator deflection at the cruise angle of attack of 1.8°. The pitching moment slope with angle of attack is shown to be negative, except immediately after stalling, resulting in the robot being longitudinally stable in flight.

The simulations in ground effect show that the lift coefficient when closest to the water surface, with the central wing being 88 mm from the water, is 28% higher than when out of ground effect, while the drag coefficient is 1% lower. This higher lift generated as the robot accelerates enables it to escape the large hydrodynamic drag from the submerged hulls. As SailMAV transitions to hydroplaning, the robot raises a further 34 mm above the water, resulting in lift and drag coefficients of 23% and 2% higher than for the cruise condition. This results in a take-off speed 5% lower that without ground effect.

11.6 Hydrodynamics

The design of the hulls has a direct impact on the performance in sailing and take-off. A poorly designed hull will increase drag to the point that take-off is hindered.

During sailing, the robot is in displacement regime, during which the entire weight of the robot is supported by its buoyancy. This creates drag not only in forward motion, but also in the lateral direction. Whilst the former is disadvantageous during sailing and take-off, the latter is beneficial for sailing as it opposes drifting. In order to take advantage of this during sailing but still achieve an efficient take-off, it is essential to design the hulls with a high length to beam ratio, the beam being the width of the hull.

A parameter that influences the transition from the displacement to the hydroplaning regime is the location and size of the step separating the forebody and afterbody of the hull (see Fig. 11.14). This step is ideally placed at the location where the coefficient of pressure is lowest, however this measurement requires complex experiments or detailed simulations with previous knowledge of the hull shape. In the pilot design presented here, the guidelines in [193] are followed and thus the step was placed 10° aft of the centre of gravity to allow the aircraft to rotate easily during take-off. The hull was then designed with a length to beam ratio of 11 with a step located at two thirds of the hull length, and an afterbody angle of 7° which limits the planing trim angle (ensuring the tail does not pierce the water interface during planing).

Figure 11.14 shows a side view of the hull with key design parameters that are determinant in the sailing, takeoff and flying performance. The forebody deadrise angle reduces motion resistance by laterally deflecting the bow wave. The scallops, generally visible as a corrugation in the forebody, further deflects the wave downwards. This diminishes the amount of spray generated, particularly important in the vicinity of wings.

11.6.1 Hydrodynamics Validation

The hull shape is validated in water tunnel experiments, where the drag is measured in function of speed, shown in Fig. 11.15. An empty hull of final dimension is held

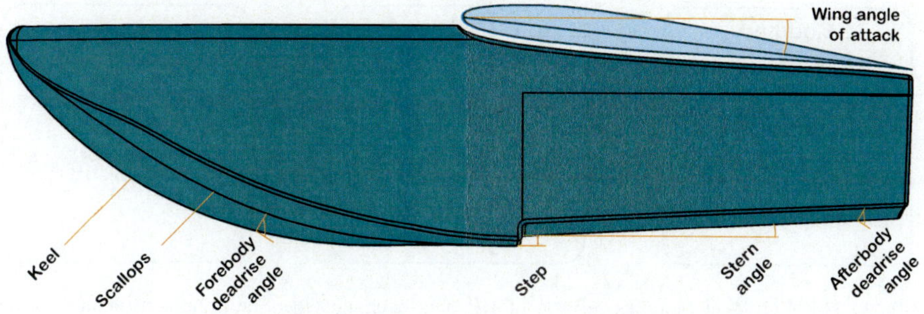

Fig. 11.14 Side view of the hull with the wing cross-section. Driving parameters are highlighted

Fig. 11.15 Drag measures of the hull in function of water speed. The deadrise angle of the keel creates the water lobes to the side, visible in this image

by a string attached to a 1D load cell while the water is flowing under the hull. The planing regime is achieved at lower velocities due to lower weight loading, as predicted by scaling laws. This enables the validation of the hulls with only a 1 m/s flow speed. Results show that hydroplaning regime is achieved below 1 m/s. Due to the experimental nature of these experiments, it is difficult to parametrically test design features, as this would be prohibitive in terms of manufacturing time. As such, this experiment serves as a final validation stage before vehicle integration, leaving the design optimisation to either an analytic approach or a simulation approach. It is important to note that both approaches are difficult due to the multi-physics (aerodynamics, moving body, hydrodynamics) aspect of this problem.

Overall, most of the design work of hulls of hydroplanes relies on tow-tank testing from 1950–1970 and a concise, up-to-date design method would be a relevant future piece of research.

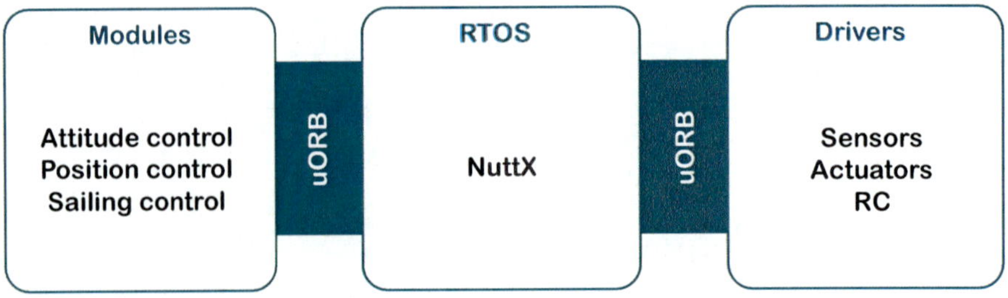

Fig. 11.16 High-level modules rely on uORB data exchange to obtain and send data with the system. The low level driver stack interacts directly with the hardware and publishes updated values to the system

11.7 Onboard Control and Autonomy

Past the 100 g mark, the payload of robots becomes large enough that off-the-shelf electronics can be a reality. Larger aerial-aquatic robot can benefit from faster development times, broader sensor choice and run more advanced software.

The motion of these robots in two media demands more sophisticated control methods to avoid uncontrolled behaviour. As opposed to gram-scale robots, falls are destructive and can even be dangerous for the operator as speeds achieved are greater. Flying autonomously in particular requires advanced control methods.

The Sailing-Flying robot presented here needs completely different control plans for the flying phase and the sailing phase. Whilst flying, a fixed-wing controlled needs to maintain stability, attitude, elevation and heading. Whilst sailing, wind direction and magnetic angle and GPS position are measured and used for the sailing controller.

This unusual setup leverages an existing flight controller platform (Pixracer) that is capable of executing the *PX4*[5] system.

11.7.1 PX4

The PX4 software is based on an advanced Real Time Operating System (RTOS), NuttX, and features a publish/subscribe messaging system for data exchange called uORB. The PX4 software is constructed with a high level of hardware abstraction, allowing for high-level control to operate independently of hardware or vehicle configurations.

The driver layer contains drivers that directly decodes peripheral signals coming from sensors, both on-board and off-board, actuators, remote control and more. The NuttX environment ensures that all drivers run according to a schedule (see Fig. 11.16).

[5] https://px4.io/.

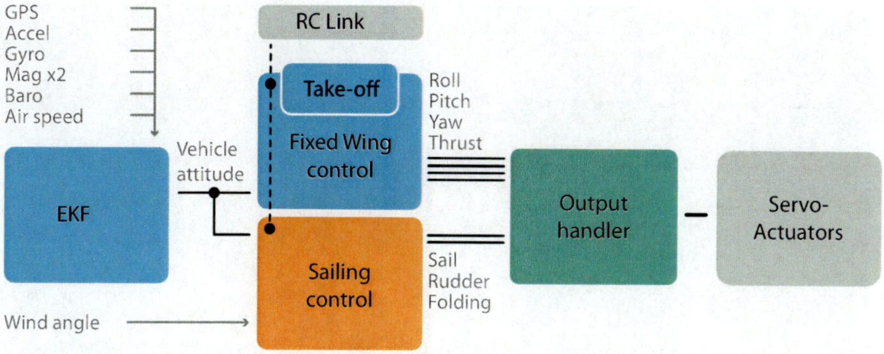

Fig. 11.17 Control diagram of the hybrid controller. The pre-existing modules from the *px4* framework are blended with a new sailing control module and an adapted output handler that serves correct values to the actuators

11.7.2 Dual-Mode Configuration

The dual configuration software demands for a method for transitioning from one state into another. The PX4 stack utilises the concept of *flight modes*. A flight mode is a specific software configuration in which selected modules run and others are stopped. Flight modes are mainly used for altering the level of autonomy in flight, e.g. from manual to stabilised flight.

The creation of a new flight mode, such as sailing, brings many advantages to the flight controller. First, the modules necessary for winged flight are not only useless in sailing but interfere by sending unwanted actuator instructions. Those modules are deactivated in sailing. Secondly, flight modes can be changed during runtime, as opposed to airframes changes which can only be realized on boot. Thirdly, flight modes can be switched from the RC link, simplifying user authority. As such, the user can decide to switch modes on demand.

Figure 11.17 shows the core control elements required for this hybrid design, detailed here.

11.7.3 Flying

The flight controller leverages existing standard fixed-wing airframe controllers. In this mode, standard controls are available, namely elevator, rudder and propellers. In addition, the two sails act as independently actuated flaps. The fixed-wing controller is available within the open-source control framework *px4*. Based on PID control laws, a dedicated module dictates how roll, pitch and yaw inputs translate into actuator commands. Take-off and landing are also handled by the standardized software. An Extended Kalman Filter (EKF) runs on a separate thread and publishes the estimated

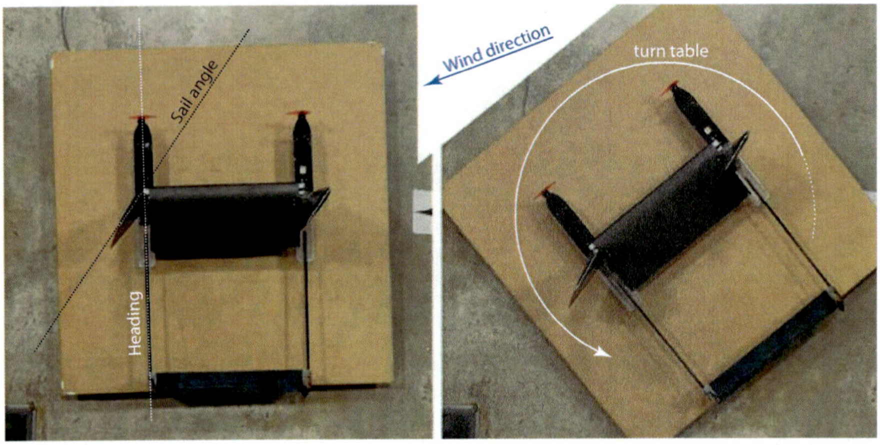

Fig. 11.18 Turn table

attitude and position based on the sensor data, i.e. GPS, double IMU, barometer and air speed sensor. Note that not all measurements obtained by the robot and shown in Fig. 11.17 are necessary for the flight controller.

11.7.4 Sailing

The sailing controller was first tested on a turn table, by positioning the robot at different angles to the wind and measuring the resulting sail orientation (Fig. 11.18). The results of this test are shown in Fig. 11.19. It can be observed that the experimental data replicates the simulated results (based on Eq. 11.5) accurately, with only minor discrepancies until the actuators physically saturate at a sail angle of approximately 60°. As a next step, the controller was tested in a water tank, using fans to artificially generate wind.

The *vehicle attitude* is estimated by the EKF and extracts the yaw angle (with respect to the magnetic north, according to a north-east-down coordinate frame). A new wind sensor measures the *wind angle* relative to the robot (Ψ_{WB}). Based on the aforementioned data, the following control law was implemented to calculate the *sail angle*.

$$\delta_{sail} = -\frac{\pi}{4}\operatorname{sgn}(\Psi_{WB})\,(\cos\Psi_{WB} + 1) \tag{11.5}$$

where Ψ_{WB} is the angle of the boat relative to the wind and is defined by the following relation, bounded by $\pm\pi$:

$$\Psi_{WB} = [(\Psi_W - \Psi_B + \pi) \bmod (2\pi)] - \pi \tag{11.6}$$

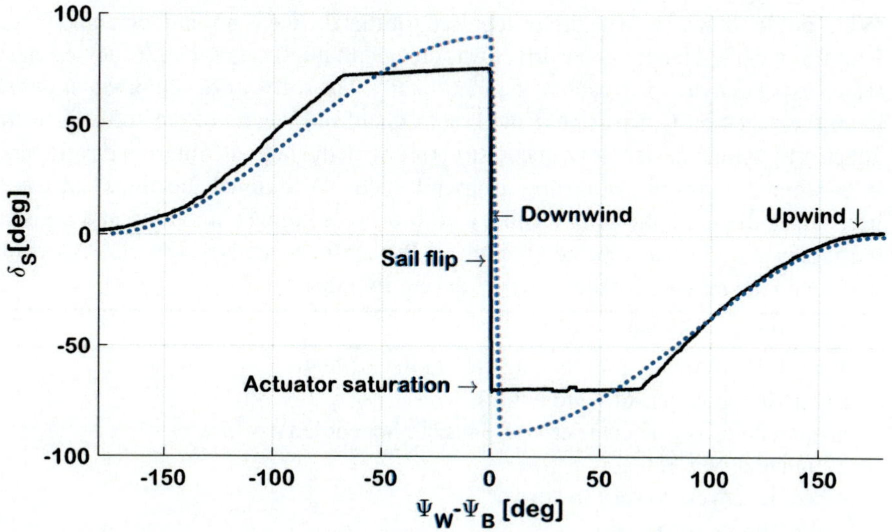

Fig. 11.19 Sail angle controller (doted line) with respect to relative wind orientation (i.e. wind-to-body angle). The solid line shows how the sail behaves experimentally: here, saturation occurs at around 65°, due to the limited servo range

Note that with the wind sensor, the angle Ψ_{WB} can be obtained directly, as long as it is wrapped to the interval $[-\pi, \pi]$. Equation 11.5 yields the sail angle with respect to the boat, and is a modification of the expression presented in Refs. [197, 198]). This formulation ensures that the sail flips from one side of the boat to the other when pointing downwind, while in the upwind boat orientation, the function goes through a continuous zero, as wanted. If sufficient actuation range is available, the sail is perpendicular to the wind when moving downwind. If a 90° angle is not achievable, as in the current vehicle, the actuators will saturate and remain at their maximum opening angle once reached (cf. Fig. 11.19 in Sect. 11.9).

First tests were effectuated without wind speed sensors. As such, the sail control law was based only on the relative orientation of boat and wind. A more advanced formulation will be developed which will take the apparent wind speed into account. While the current sailing control scheme is basic, it allows for autonomous sail orientation, thus enabling the boat to exploit the available wind as best possible, and highlighting one of the key capabilities of the developed robot. Note that the sail is adjusted exclusively to fulfil this purpose, while the heading is adjusted using the rudder.

Future work will extend the functionalities of the sail controller to include autonomous path-following, as well as additional modes to permit the boat to move upwind with a tacking strategy, to turn downwind without losing speed via a gybing manoeuvre, and to move in the absence of wind, exploiting the propellers.

Wind angle software The driver required for this device is not implemented in the PX4 framework. Hence, a new driver was created that leverages the I2C device class which exposes byte-wise read-write functionality. The software development is based upon the excellent C driver published by SOSandroid.[6] Parts of the software were adapted to use the newer work queue structure with the help of Auterion developers. Algorithm 1 shows the work flow in pseudo-code. Additional functions not listed here are available in the source code, e.g. to provide signal smoothing and register burning capability. The repeated reading of the sensor is managed by the PX4 work queue and the reading frequency is set during initialisation.

Algorithm 1: AS5048 driver

```
Result: Periodically poll the angle from the AS5048 IC
advertise topic (sensor wind angle)
instantiate driver object (derive from I2C, ScheduledWorkItem)
initialize driver, I2C
schedule a cycle to start the driver
while Driver started do
    function run:
        function collect:
            request data from sensor over I2C
            convert data to angle
            publish angle to topic
        schedule a new loop at set frequency
end
```

11.8 Operation of a Sailing-Flying Robot

The SailMAV robot was tested in sailing, powered acceleration and in take-off. The transition to flight in particular underwent rigorous testing both indoors and outdoors, in mild conditions. The aircraft was able to successfully navigate passively on the surface in cross-winds and take-off in about 10 m. These results are discussed in the next sections. They show a solid foundation towards full autonomous deployment but further testing will be required to ensure reliable operation, a better understanding of the performance of the novel robot and optimisation. This is planned in the near future and subsequent findings considered for publication.

[6] github/sosandroid.

Fig. 11.20 Flight-sailing-flight cycle overview. The robot lands on water where is transitions to a *sailing* configuration and executes its mission. Take-off is achieved via propelled acceleration, transitioning from displacement vessel regime, to hydroplaning and lastly to being fully airborne

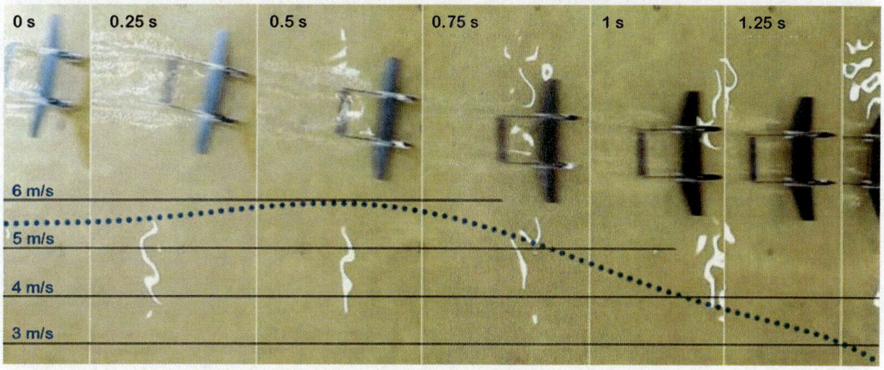

Fig. 11.21 Top-view and velocity profile of the robot reaching peak velocity in the available space

11.8.1 Transition from Water to Flight

Hydroplaning was tested in a shallow water tank (10 cm depth), in calm, controlled conditions, and was achieved within 5 m, at full throttle with a wide hull shape. Figure 11.21 shows a top view of the final part of the trajectory in one of these tests, where a maximum velocity of 6 m/s was achieved, slightly below take-off speed. The tank size limits higher velocities.

Outdoor experiments were also performed to assess the robot's transition from water to flight, as shown in Fig. 11.20, which presents the different phases of the transition. In the displacement regime, the robot is moving slowly and the hulls are fully in the water. It can also be seen that the tail is submerged during the moments before transition to hydroplaning, which creates spray and considerable drag. This is however not the case at lower speeds or during hydroplaning. Increasing the beam of the hulls was found to help reducing the spray towards the propulsion system, which prevented the robot from accelerating effectively for lower beam lengths.

11.8.2 Sailing

Figure 11.22 shows a cross-wind sailing manoeuvre. In this test the rudder is fixed to a given angle. The robot is autonomously able to maintain its heading and move crosswind effectively.

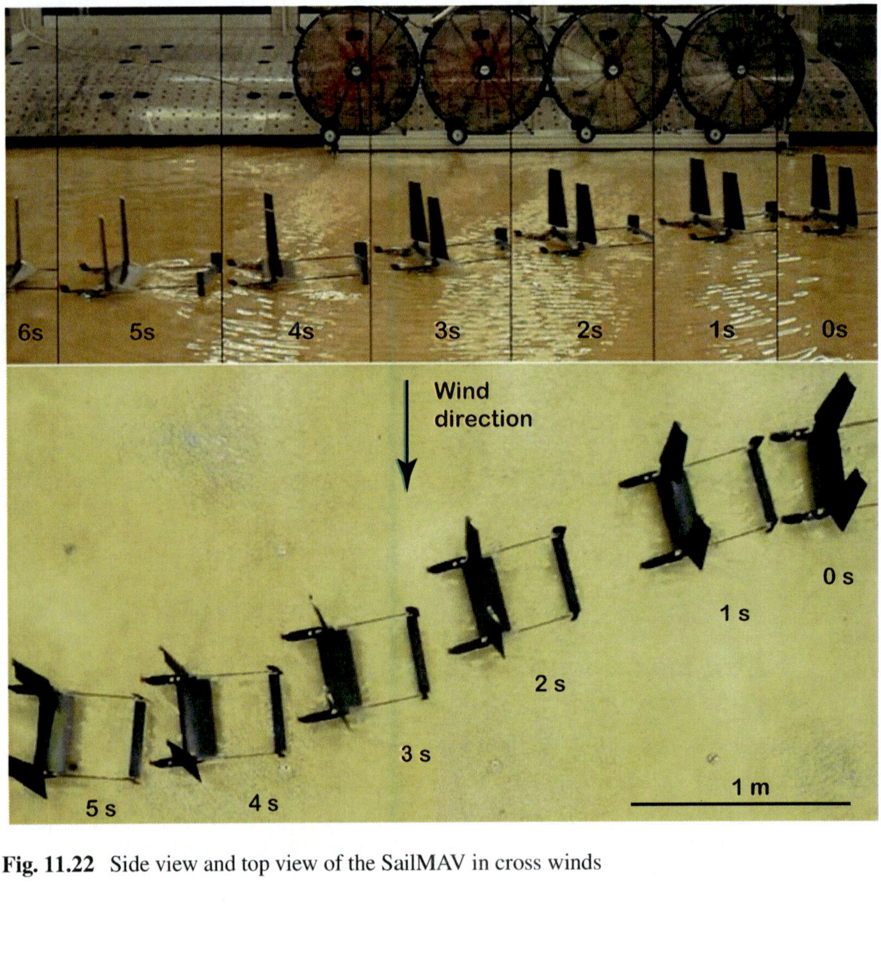

Fig. 11.22 Side view and top view of the SailMAV in cross winds

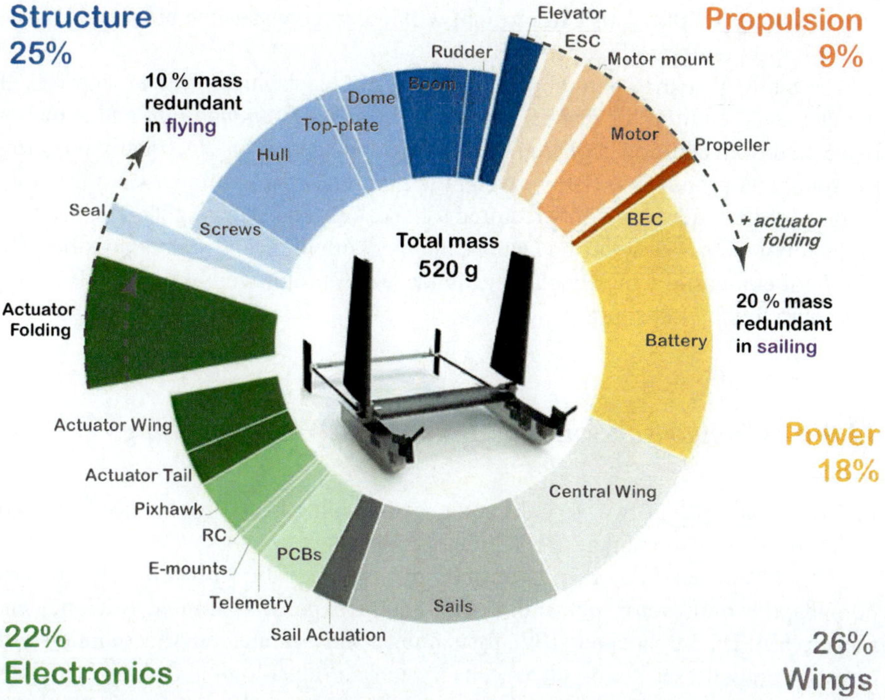

Fig. 11.23 Mass distribution of SailMAV, split into five categories: structure, propulsion, electronics, power and wings. The exploded bars show the *penalty* associated with the hybrid design in both sailing and flying

In sailing, two turns exist, referred to as *tacking*. Tacking generally implies turning upwind, where the boat's prow points towards the wind. There also exists another turn, the downwind turn, called *gybe*. Manual tacking with the robot is a demonstrated operation. However, simultaneous control of the rudder and sail angle manually, via remote control is difficult for an untrained pilot. Future work should integrate that manoeuvre into the controller. *Gybing*, is a more robust turning manoeuvre as the robot does not lose speed and as such, rudder control. However, this comes at the cost of downwind drift.

11.9 Characterisation

One of the fundamental issues with hybrid designs is the mass penalty incurred by the different modes. The developed robot has 10% redundant mass in flight, linked to the waterproofing and the folding of the wings, both of which are unnecessary in flight (Fig. 11.23). In sailing, the entire propulsion block, including motor and drivers is redundant. The folding of the wing element is also irrelevant, bringing the penalty during sailing to 20%. This extra weight will lower the water line but should have a minor impact on the sailing performance.

The SailMAV carries a total of 22% in mass of electronics, which comprises all the elements required for manual as well as fully autonomous flight and actuation in all modes. In addition, a quarter of the weight is needed for electrically powering the robot and propulsion. Battery weight is minimal, representing 14% of the total weight, as the batteries are only required for take-offs and short flights. The last 50% of the robot mass is structure and wings, making both flight and sailing possible. The ultralight composite wings could be extended with minor weight penalty, lowering minimum flight speeds further.

11.9.1 Comparison with the AquaMAV: Wing Morphing

In order to operate in two very different media, animals and robots need to adapt their morphology to maintain good efficiency levels.

While there has been demonstrations of aerial-aquatic operation of standard, non-adaptive multi-rotor robots, this comes at the expense of limited efficiency and range [160]. The Looncopter [109], for example, shows underwater locomotion with the same propellers as for flight, as well as water-air transitioning, making it applicable for short, fast and flexible missions. In addition to the short lifetime and range of multirotors, it has been shown to take off only in flat waters. Fixed-wing designs [117] improve on the aforementioned limitations, but typically do not show mobility on the water surface or underwater.

The changing morphology allows both of these platforms to operate more efficiently. Wing morphing in the AquaMAV allows it to dive passively and reach a depth of 3 m, which, besides reducing drag in the water, saves redundant weight in flight, by not requiring a buoyancy control system (Fig. 11.24). Moreover, the SailMAV's morphing wing allows it to fly without a fixed vertical sail, which would generate a large off-centred drag component and impact flight stability. Morphology changes bring substantial benefits in terms of weight, compared to single-configuration robots. Table 11.5 shows the penalty that the hybrid designs incurs.

This was estimated by measuring individual component mass relevant to each mode. In addition, the hybrid design requires adaptations that impact performance. For the AquaMAV, for instance, a zero-thickness wing helps the robot to survive the plunge-dive and stay neutrally buoyant. For the SailMAV, a symmetric profile is

Table 11.5 Measured redundant weight in air and water

Weight penalty	Water (%)	Air (%)
AquaMAV	29	20
SailMAV	20	10

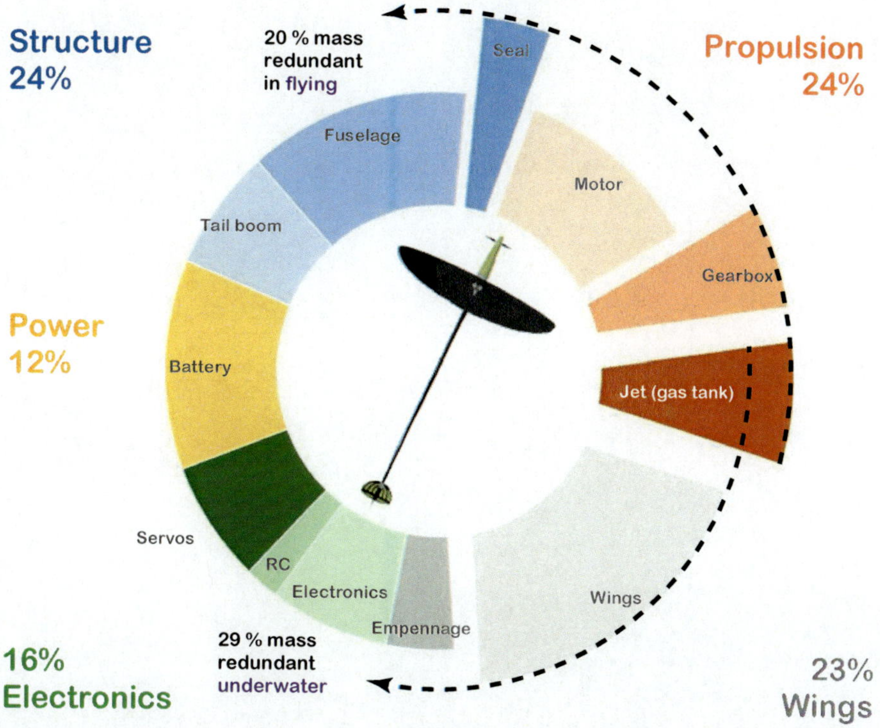

Fig. 11.24 Mass distribution of AquaMAV, split into five categories: structure, propulsion, electronics, power and wings. The exploded bars show the *penalty* associated with the hybrid design in both underwater and in flight

used for the wings, to obtain the same behaviour when sailing under port or starboard wind. Both these design features necessary for in/on water operation lead to a reduced efficiency in flight.

Chapter 12
Multirotor Aircraft and the Aquatic Environment

The previous chapters presented hybrid robot concepts and prototypes relying on the use of fixed wings for lift generation. The higher flight efficiency of such devices makes them suitable for covering large distances and can even serve to extend their locomotion envelope (see Chap. 11). However, this technology choice has two important implications. Firstly, the complexity of landing and take-off is substantially increased due to the forward speed required to perform these manoeuvres with fixed-wing robots. This drawback can be mitigated by relying on impulsive take-off and diving manoeuvres, elegantly eliminating the need for horizontal acceleration/deceleration phases, however the impulsive approach also comes with penalties (see Chaps. 6 and 9). Secondly, the wings mostly become a nuisance underwater. Their large size negatively impacts drag, even if some form of wing folding is used to attenuate this problem. The wings also need to be neutrally buoyant, forbidding the usage of thick wing profiles (airfoils, as opposed to flat plates).

Multirotors are a type of platform that can also tackle aerial-aquatic environments, circumventing the above-mentioned drawbacks. They have become ubiquitous in recent years, through rapid advances in sensor technology as well as lightweight computing boards capable of stabilising the robots and efficient brushless motor drive electronics. Nonetheless, the use of multirotors for aerial-aquatic missions also presents significant challenges, as discussed here.

12.1 Introduction to MEDUSA

There already exists various aerial-aquatic multirotor solutions, most of which have been successfully implemented and tested (see Sect. 4.4). However, these systems have limitations, in terms of underwater efficiency and communication. Indeed, wireless signals can only penetrate to a shallow depth while drawing significant power to overcome absorption.

R. Zufferey et al., *Between Sea and Sky: Aerial Aquatic Locomotion in Miniature Robots*, Biosystems & Biorobotics 29, https://doi.org/10.1007/978-3-030-89575-4_12

Rather than designing a single robot capable of operating efficiently in both environments, one alternative is to use a *pair* of robots inspired by the symbiotic relationships sometimes found in nature.

Many animals develop such relationships to overcome their natural morphological shortcomings. An example is the relationship between remoras and sharks. By attaching itself to a shark, the remora is able to travel long distances while being protected by the shark and expending only minimal amounts of energy. At the same time, the remora eats bacteria or parasites that have grown on the shark's skin, which helps the shark avoid infections. In nature, symbiotic relationships develop out of a need for food, protection and/or hygiene. While robots face different challenges, the same principle can be applied, where the respective strengths of two robots compensate for each other's weaknesses.

One solution that has been suggested, in a somewhat similar vein, is the use of low-flying quadrotors dropping rope-attached sensors into the water and then retracting them again [199]. This is a relatively robust approach but is constrained by the short battery time characteristic of quadrotors. This implies that the collection of a sample or measurement must happen quickly, and repeated few times. Moreover, no accurate control is provided over the positioning of the dropped sensor in the water.

In the remainder of this chapter, we discuss a new solution for robotic water sampling, comprising a submersible pod tethered to a quadrotor. The quadrotor flies to a designated location, lands on the water surface and deploys a pod to collect measurements. Landing on the water surface, as opposed to hovering above it, allows for longer-duration missions when compared to existing aerial robot-based systems, as the quadrotor can float passively while the pod performs measurements. When compared to single-robot systems, the advantage of our system lies in its modularity and ease of development. The quadrotor is based on off-the-shelf components, so it does not need to be designed and manufactured from scratch, and remains outside the water, hence requires little waterproofing. Moreover, single-robot systems designed to collect data underwater still face challenges with demonstrating air-water transition, controlled flight and controlled underwater locomotion in a single system [3].

This chapter presents four novel contributions to the field of aerial-aquatic locomotion. Firstly, we demonstrate a novel approach for robotic water sampling, that is based on a conventional multi-rotor platform and better suited for rapid field testing than available methods. Secondly, we discuss the development of communications, electronics and control frameworks necessary for a symbiotic dual-robot system. Thirdly, we present the integration of a new prototype (MEDUSA: Multi-Environment Dual-robot for Underwater Sample Acquisition, shown in Fig. 12.1) which validates the proposed concept. Lastly, the system is shown in operation in both a lab environment and outdoor field tests.

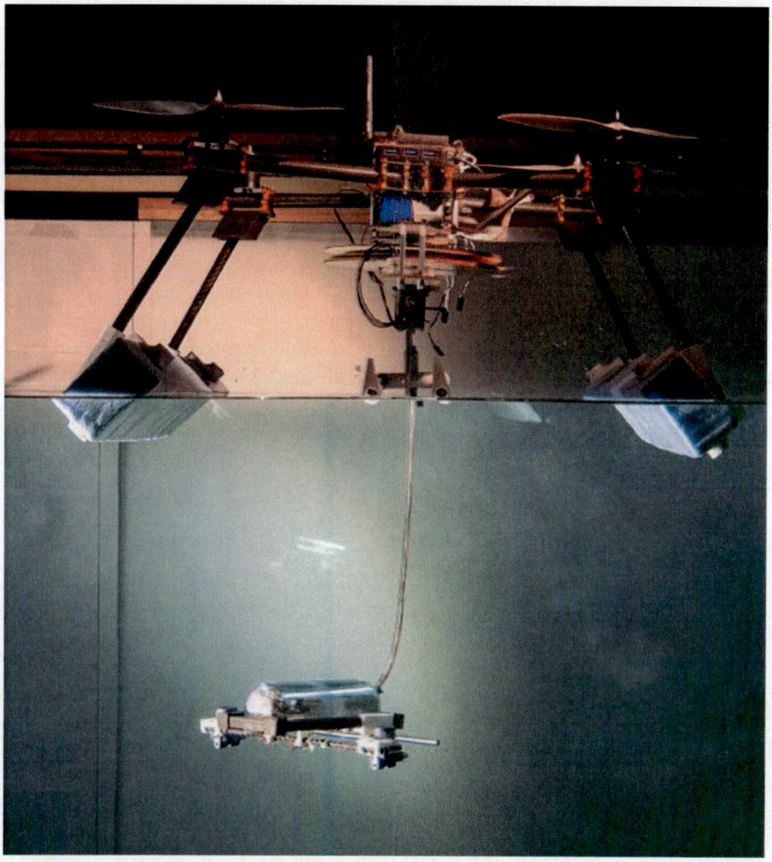

Fig. 12.1 Photo of the MEDUSA system in action: the multirotor robot is floating, while the pod has been partially uncoiled and is moving underwater

12.2 Design

12.2.1 Operation Principle and Envelope

The MEDUSA symbiotic robot system is specifically designed for water sampling in difficult-to-reach locations. As shown in Fig. 12.2, the system consists of a quadrotor that can fly to specific locations, land on the water and deploy a smaller robot which collects water measurements or samples. The smaller robot (hereafter 'underwater pod') has 3 controlled degrees of freedom. It controls its depth through inflation of a soft membrane, its forward motion via the thrust produced by two pumps on the outside of its body, and its orientation around the vertical axis via a thrust differential between the same pumps. This allows the system to travel successfully to different difficult to reach and potentially far apart locations, and to collect various

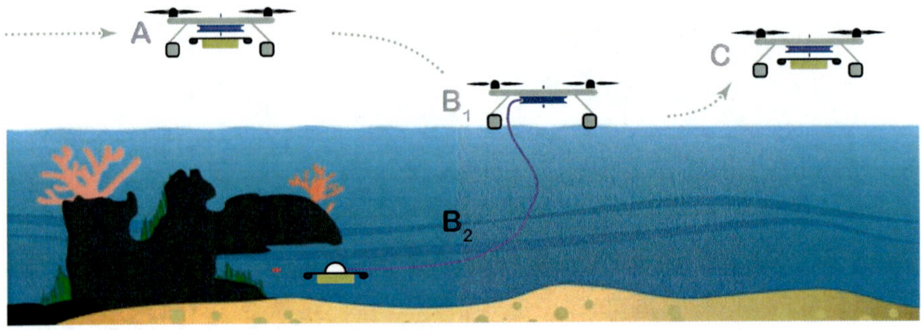

Fig. 12.2 Schematic of the operation principle of the dual system. **A.** The flying module carries the pod (yellow) to set location. **B1.** The flying module lands on the water surface and floats due to buoyancy elements. **B2.** The coiler subsystem (blue) releases the tether and the pod moves vertically (buoyancy control) and horizontally (jet) to reach the target underwater location. Live video is transmitted and a sample taken. **C.** The tether is coiled again and the drone flies back to the user

water samples or measurements within a single mission. The pod is connected to the quadrotor by a 5 m long tether, which constrains its operating envelope to a spherical cap with slightly under 5 m radius.

A specific scenario where MEDUSA can operate and perform better than other aerial aquatic solutions is the monitoring of coral reefs, or similar fragile underwater ecosystems. These environments are very unstructured and normally encountered at shallow depths. The five-metre range of the tether is thus sufficient for the mission and the underwater pod can manoeuvre the intricate features of the reefs due to its small form factor and nimbleness. Thanks to the features mentioned, water samples can be obtained in the vicinity of relevant biomes that would otherwise not be possible to access remotely.

12.2.2 Deployment Flying Vehicle

The flying platform is a custom-built quadrotor of 1 kg and a payload of 800–900 g. The double 2D plate frame is well-suited to affix the bottom-mounted coiling system, as shown in Fig. 12.4.

The underwater mobile pod depends entirely on the *tether* for power, commands, relaying back data, live video and air supply. A design constraint for this system is its maximum depth, which has a direct impact on the required tether length. In order to have an operation envelope of a spherical cap of 5 m radius, an equivalent length of tether needs to be reliably coiled before flight and uncoiled once in the water. Furthermore, in order to minimise the impact of the tether on the pod motion, its stiffness needs to be reduced as much as the application permits and its buoyancy must be close to neutral. The developed custom tether, shown in Fig. 12.4 (bottom), consists of a 6 mm outer diameter silicone tube that shields the cable from sea-water and is bonded at its extremities. It contains a curved ribbon cable by Cicoil. Every strand is made from a twisted copper wire with silicone insulation, chosen with a 1 A specification. The ribbon cable leaves an air gap within the silicone tube that serves as a channel for air flow. The copper strands limit the minimum bending radius, permitting unrestricted air flow at all times. The end of the tube is sealed onto the air pump resting in the pod.

Underwater Pod The underwater pod consists of two external pumps, which steer the pod via differential thrust, and a main cabin, which houses the electronics, pump and membrane needed for vertical locomotion. This pump is sized to withstand pressure 5 m underwater and determines the pod achievable depth. Inside the main cabin is a microcontroller, which handles communication between the quadrotor and the underwater pod. The cable that connects the underwater pod to the quadrotor via the coiling unit is used to send high-level commands to the pod. The actuators, sensors and microcontroller inside the pod are also powered by the quadrotor, with the wires connecting these systems to the quadrotor's battery also passing through the coiling mechanism. The main cabin is pressurised, reducing structural loads, therefore reducing weight and enhancing waterproofness.

The structure of the pod is shown in Fig. 12.3A. Its dimensions are minimised for the smallest volume possible in order to reduce the amount of ballast needed to sink. For that purpose, the shape of the pod was designed to match the containing components. A semi-cylinder fits well and rests against a rigid support plate.

The PETG dome is vacuum-formed and has a lip to seal against the PMMA support plate as shown in Fig. 12.3B. The silicone seal is made from Moldstar 15. It is maintained under pressure through a 3D-printed nylon clamp. The 4 C-Clamp segments fit against each other to seal the entirety of the perimeter. Rubber bands in the corners lock the clamps in place. This sealing method is reliable (more than 100 opening cycles), takes 10 s to seal and can be rapidly prototyped.

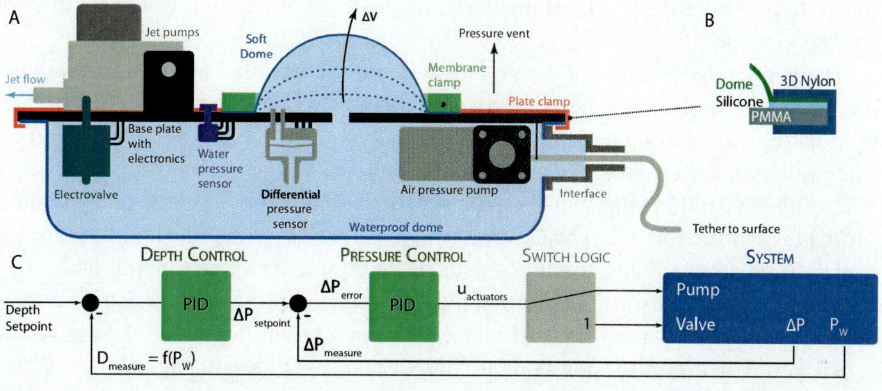

Fig. 12.3 **A.** Side view of the underwater pod with its main internal components. **B.** Seal clamp structure. **C.** Depth control schematic

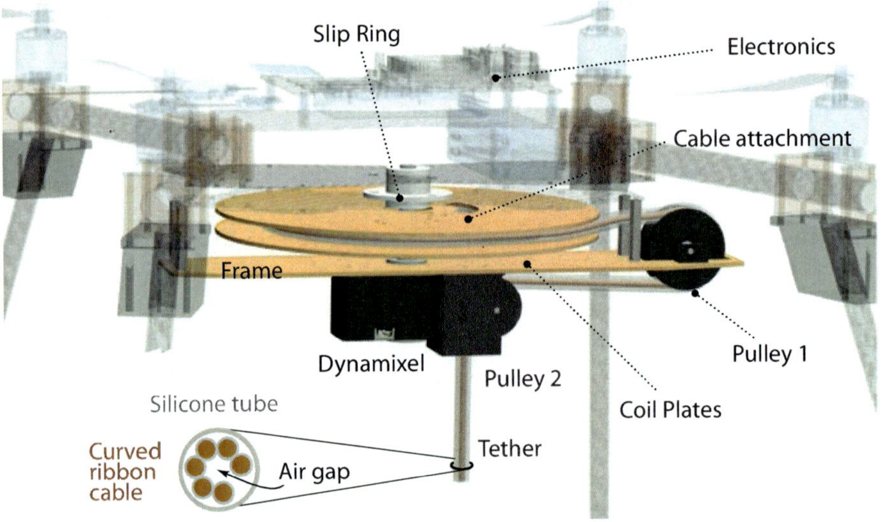

Fig. 12.4 CAD view of the coiling mechanism installed at the base of the X-frame quadrotor

12.3 Sensing and Control

12.3.1 Horizontal Actuation and Buoyancy Control

The orientation of the underwater pod is controlled by two external jets which steer via thrust differential. When the floating quadrotor receives a command to turn the underwater pod, it adjusts the thrust generated by both of the external jets in order to rotate the pod. In the original implementation of the robot, this is done via open-loop commands, however a closed-loop control scheme could be easily implemented and integrated within the existing framework, if a specific orientation were required for a given mission.

The vertical locomotion of the underwater pod is controlled via buoyancy adjustments. This strategy is inspired by fish, who generally control their depth in the water by inflating or deflating a bladder inside their body, thus changing their buoyancy. The presented underwater pod employs the same technique to control its buoyancy and resulting depth, using two pressure sensors and a pump. The buoyancy control process is clarified in Fig. 12.3C. Once the pod receives a depth setpoint from the user, it uses an absolute pressure sensor to determine its current depth and a differential pressure sensor to determine the difference in pressure between the inside and outside of the cabin. The difference between current depth and desired depth is fed into a first control component ('depth control' block in Fig. 12.3C), which determines the differential pressure required for the pod to attain and maintain the target depth. The computed pressure request is monitored to ensure that it remains within achievable bounds and does not cause the membrane to burst. A second PID controller ('pressure control') takes the difference between the *measured* differential pressure and the aforementioned *required* differential pressure as error signal, and computes the required actuator inputs.

While a single signal is computed by the controller, the pod is controlled by the combination of a pressure pump and a valve. In a final step, the input to each actuator is therefore determined according to the following logic. If the actuator signal computed by the controller corresponds to an *increased differential pressure demand*, the signal is passed directly to the pump and the valve is closed, allowing pressure to build up. If, on the other hand, a decreased differential pressure is required, the pump is set to zero and the valve is opened to allow for pressure to be released.

12.3.2 Sensor Payload

The MEDUSA robot is equipped with a camera and an absolute pressure sensor, so that underwater images and their corresponding depths can be recorded. The camera is physically placed in the pod, it thus transmits underwater footage when the pod is deployed and aerial footage (fixed heading) when in flight. In future iterations, these sensors could be complemented by salinity, pH, oxygen content and other water quality sensors, as required by the specific mission.

Research into micro filtering systems is currently ongoing. The ability to take multiple samples at various depth is one of the requirement for biologists. This gives researchers the ability to measure chlorophyll content in shallow waters. This metric is particularly important to assess chlorophyll blooms. Initial results show that such a device can be assembled and fitted into MEDUSA's pod. The design consists of a micro peristaltic pump which pumps a known volume of water through a paper filter. The particles get stuck on the filter which is analysed after the mission. A set of valve fitted to the pump adds additional filters, for multiple sampling.

12.3.3 Electronics, Communication and Inter-System Tether

The MEDUSA drone comprises a uniformed electrical network that spans from the flying vehicle to the underwater mobile pod (cf. Fig. 12.5). The companion computer runs ROS (Robotic Operating System), which handles high-level tasks and all the control of the system. It interfaces with the *coiler* subsystem, in charge of coiling and uncoiling the tether. The coiler is operated via a dynamixel servoactuator, running as a node in ROS. Two different serial links span from the companion microcontroller to the flight controller (MAVLINK, Micro Air Vehicle Link serial communication protocol) and to the underwater pod using ROSserial.

The flight controller handles the framework required for multirotor flight. It directly outputs the signals to the four Electronic Speed Control (ESC) board that

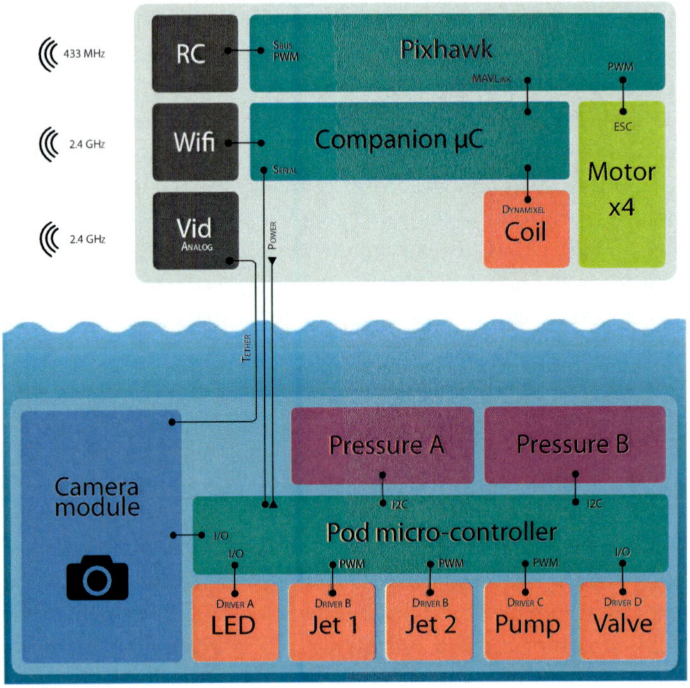

Fig. 12.5 Communication and electronics framework showing both the drone subsystem (top) and the pod subsystem (bottom), linked by a custom tether

drive the brushless motors. Additionally, the flight controller relays flight data to the *companion computer* via MAVLINK.

The end-user interacts with the *flight controller* via a standard 433 MHz radio link, which provides kilometre-range operation. This connection transmits instructions and manual control commands and is supplemented by a telemetry link. The latter serves to send and receive commands that can change parameters and read system states. Analog video is broadcast by radio and requires only one cable in the tether.

12.4 Results and Discussion

The developed system was tested both in a controlled laboratory setting, allowing for extensive underwater characterisation, and in realistic outdoor conditions. The experiments aimed to demonstrate the locomotion and manoeuvring capability of the system in both the horizontal and the vertical plane, the effectiveness of the depth control system, and the overall functioning of the robot in realistic outdoor conditions. Furthermore, preliminary tests were performed to assess the efficacy of the membrane-based buoyancy control method proposed, and a model was developed

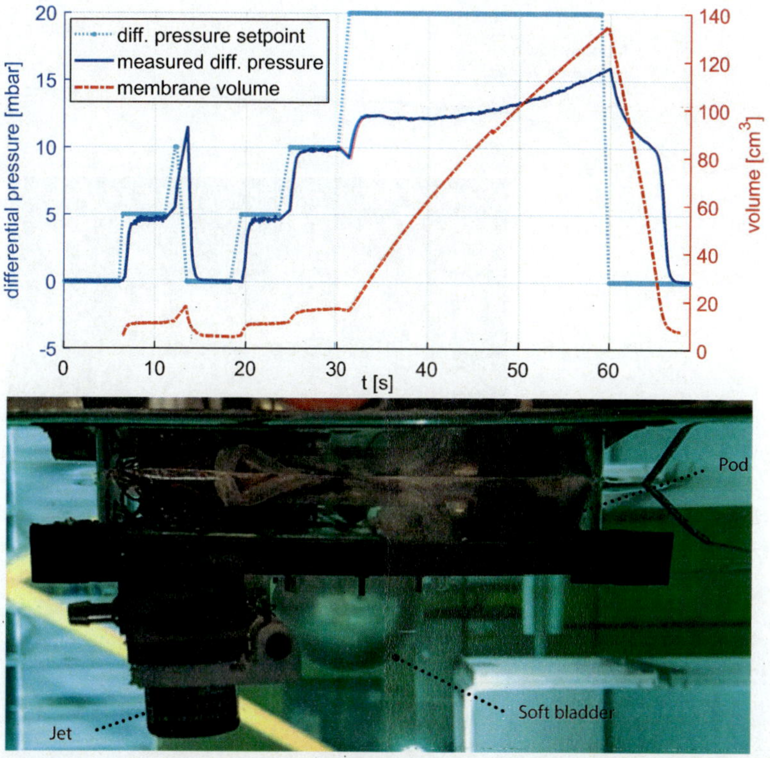

Fig. 12.6 Evaluation of the buoyancy-based actuation mechanism for vertical motion. **A**. Measured membrane volume versus differential pressure for different pressure setpoints. **B**. View of the inflated membrane during experiments

to yield more detailed insight into the behaviour of the membrane. The remainder of this section provides an overview of the main results obtained.

12.4.1 Membrane-Based Actuation

As elucidated in Sect. 12.3.1, the vertical motion of the underwater pod is controlled through pressure-based inflation and deflation of a soft membrane (as shown in Fig. 12.6B). An important preliminary step is to establish whether this type of actuation mechanism is adequate for stable depth control. To evaluate this, we fed the pod a number of differential pressure setpoints and observed the resulting changes in differential pressure and volume of the deforming membrane. This is equivalent to testing the system with only the inner loop 'pressure control' component of Fig. 12.3C active. The aforementioned experiments were conducted outside the water with the pod at rest, so as to provide a clear view of the inflating membrane. The volume of

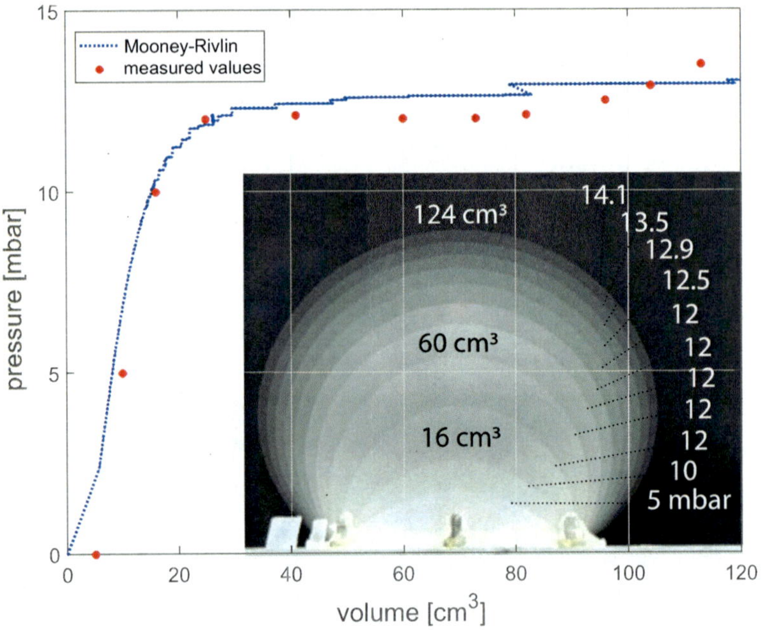

Fig. 12.7 Membrane volume versus pressure: model-predicted and tracking data-reconstructed results

the membrane at each time step was computed based on video footage. The inflated membrane is approximated as a spherical cap, whose volume can be computed from its base width $2a$ and maximum height h via the expression,

$$V = \frac{1}{6}\pi h(3a^2 + h^2). \tag{12.1}$$

Subsequent time synchronisation with the concurrent sensor measurements provided the corresponding differential pressure in the pod. As anticipated, and clarified by Fig. 12.7, the volume of the soft membrane increases as pressure is increased. The two variables are approximately proportional up to a differential pressure of 10 mbar, after which point the relation between the two becomes significantly nonlinear, with a strong increase of volume at nearly constant pressure. It can also be seen that there is a limit to what differential pressures can be reached. Based on both observations, it was decided to limit the differential pressure to values below 10 mbar, where the membrane is not in danger of bursting and the relation between pressure and volume is approximately proportional. Within this region, the pod can be rapidly stabilised at any given pressure (cf. Fig. 12.6A).

In order to further understand the behaviour of the membrane, it was modelled as a thin, hyperelastic, incompressible shell using the formulation developed by Adkins and Rivlin [200]. The membrane is described using a cylindrical coordinate system (r, Θ, z), where each particle on the surface of the membrane exists at a point (ρ, θ, η). The stretch ratios in the meridional and circumferential directions are λ_1 and λ_2, respectively.

$$\lambda_1 = \sqrt{\frac{\left(\frac{\partial \rho}{dr}\right)^2 + \left(\frac{\partial \eta}{\partial r}\right)^2}{1 + \left(\frac{\partial z}{\partial r}\right)^2}} \tag{12.2}$$

$$\lambda_2 = \frac{\rho}{r} \tag{12.3}$$

The equilibrium equations of the deformed membrane yield

$$\frac{\partial T_1}{\partial \rho} + \frac{1}{\rho}(T_1 - T_2) = P_t \tag{12.4}$$

$$K_1 T_1 + K_2 T_1 = P_n \tag{12.5}$$

where T_1 and T_2 represent the meridional and circumferential stresses, P_n represents the stresses normal to the membrane surface and P_t represents pressure in the transverse direction. For this problem $P_t = 0$. Finally, assuming that this is an incompressible material that can be modelled with a two term Mooney-Rivlin constitutive relation, we get

$$T_1 = 2C_1 h \left(\frac{\lambda_1}{\lambda_2} - \frac{1}{\lambda_1^3 \lambda_2^3}\right)(1 + \alpha \lambda_2^2) \tag{12.6}$$

$$T_2 = 2C_1 h \left(\frac{\lambda_2}{\lambda_1} - \frac{1}{\lambda_1^3 \lambda_2^3}\right)(1 + \alpha \lambda_1^2) \tag{12.7}$$

where C_1 and α are constants that come from the constitutive relationship. This relationship is usually written in terms of C_1 and C_2 where $\alpha = C_1/C_2$.

The above equations were simplified into three first-order ordinary differential equations (ODEs) using the method described by Yang and Feng [201] and then solved using a Runge-Kutta algorithm, implemented as in Ref. [202]. Solving these ODEs yields the height of the inflated membrane bubble, which can then be used to calculate the volume as per Eq. 12.1. The material properties, C_1 and α, could not be found for our particular material (Ecoflex 00-20), so values for a similar material were chosen as a starting point, and they were then adjusted to fit our data [203]. The values used to describe the geometry of the membrane and the corresponding material constants came from fitting the model to the data.

12.4.2 Vertical Locomotion and Control

The first set of indoor tests was performed in a 1 m deep water tank, in order to evaluate the effectiveness of the depth control scheme. These experiments involved placing the pod in the water, and successively sending it different depth setpoints to track, while always remaining attached to the quadrotor via the tether. An example of the obtained results is illustrated in Fig. 12.8, where it can be observed that the pod is able to track user-defined depths effectively, and that the developed buoyancy-based actuation can be used for vertical manoeuvring. New setpoints are consistently reached within approximately 5 s—a short time, given the relatively slow motion of the pod in the water.

Further it was found that, especially when moving downwards (to greater depths), the pod develops a slow oscillatory movement around the target depth. This may indicate that the actuation mechanism is unable to fully cancel out all oscillatory motion. Nonetheless, the observed fluctuations typically have a magnitude of approximately 2–5 cm. Considering the intended domain of operation in the range of several metres, as well as the non-ideal experimental conditions, which inevitably involved some water movement, the attained accuracy of the system can be considered substantial despite the aforementioned oscillations. Note that, by contrast, when moving to smaller depths, the robot appears to stabilise more accurately and rapidly. This difference in behaviour is likely due to the combined pump and valve-based actuation explained in Sect. 12.3.1, according to which increases in depth are obtained through valve opening.

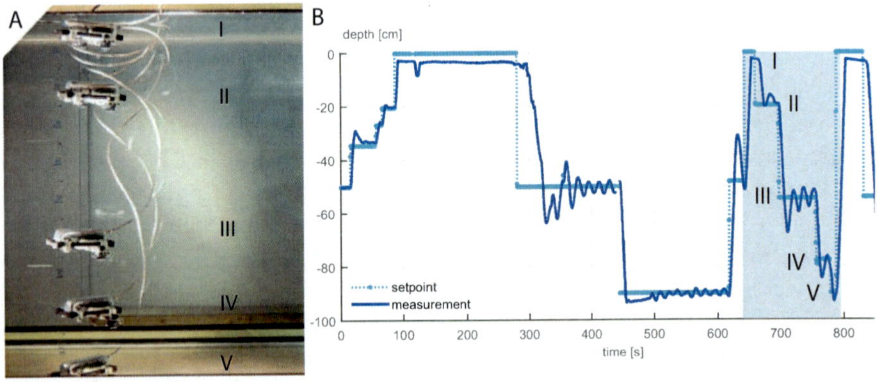

Fig. 12.8 Vertical actuation and depth control results, showing the pod tracking a number of different successive depth setpoints

12.4.3 Horizontal Locomotion

While large horizontal distances are covered by displacing the quadrotor itself, and horizontal movement of the pod can hence be limited in range, it is an important asset that the pod is able to manoeuvre and position itself within the region of interest. It may for instance be necessary for a particular mission to approach underwater objects, ranging from oil platform risers to coral reefs, or even to move underneath or inside a structure such as a pipe. In this context, the lateral manoeuvrability and performance of MEDUSA was tested.

Results show that the jet-based actuation provides sufficient thrust, leading to a manoeuvrability well-suited to the types of missions envisaged for the robot. As shown in Fig. 12.9 (left), the pod can reorient itself rapidly and perform horizontal turns at up to 100°/s, as well as reach given target locations within the range allowed by the tether, reaching a speed of up to approximately 0.35 m/s (cf. Fig. 12.9).

While the actuation mechanism does not provide direct control over all degrees of freedom, the combination of vertical and horizontal actuation leads to a comprehensive range of motion. By manoeuvring the pod into a closed space with a single lateral

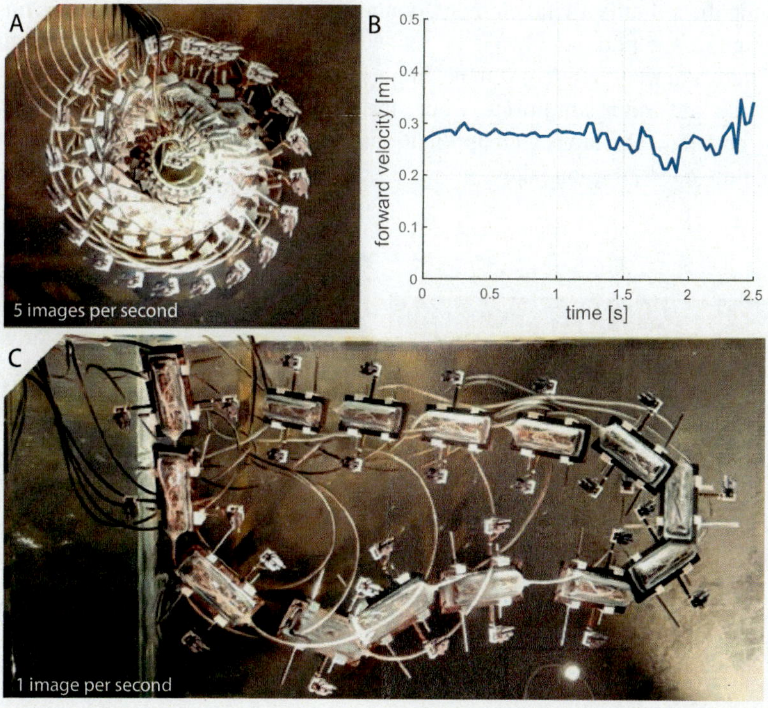

Fig. 12.9 **A.** Composition showing maximum rotational control of the robot at 100°/s. **B.** Manual control of the horizontal motion of the robot moving forward, turning and returning to base. **C.** Maximum velocity of the robot with jet thrusters at 100%

opening (cf. Fig. 12.10 left), we for instance showed that narrow confined spaces can be reached, which is an important capability for the type of system proposed.

12.4.4 Outdoor Testing

Outdoor field tests were performed on a lake, surrounded by dense forest (cf. Fig. 12.10). The water conditions were calm, but there were gusty winds. One of the most critical aspects of an aerial-aquatic robot is the transition from air to water and vice-versa. We successfully demonstrate multiple landings and take-offs from the water of the quadrotor carrying the coiling unit and pod. The downwash of the propellers does not create significant spray during take-off and in general no significant difficulties were encountered. The tests suggest that the robotic concept can function reliably and robustly in realistic and challenging conditions, similar to those that would be encountered in a typical application scenario. Underwater operation of the pod is not shown in the reported field tests due to the low visibility in the pond (less than 10 cm). This would have hindered any visual observation of the position of the pod or any relevant remote video recording.

While the tether is a vital element, containing electronic and pneumatic links and ensuring that the pod can always be recovered, it also limits the operating range of the system and may tangle or break. In general, MEDUSA is designed to operate in mild wind and water conditions, where the quadrotor can float in an approximately steady position. The envisaged applications would take place in favourable weather or in calm waters, e.g. on lakes.

Fig. 12.10 **(Left)** The robot is remotely manoeuvred into a confined space with a side opening of 30 cm **(Right)** The flying dual-robot platform shown during a take-off for a return to the user. Location: Silwood Park

12.5 Conclusions

Following an overview of multirotor aerial-aquatic locomotion approaches, this chapter presented a novel concept for automated water sampling in difficult-to-reach locations, based on a dual-robot system consisting of a multirotor that can reach remote locations rapidly and land on the water, and an underwater pod that can stream live video and take samples. The proposed solution significantly reduces the limitations and difficulties typical of single-body aerial-aquatic robots, while still benefiting from the same advantages. The development of a prototype demonstrating the approach has also been discussed, as well as the design trade-offs, implementation and manufacturing methods required to build and operate such a system. Indoor and outdoor tests showed that the approach can be applied successfully in realistic conditions, and that the soft membrane-based depth controller and jet-based horizontal actuation allow for accurate positioning and manoeuvring of the pod. The proposed approach allows for challenging underwater missions to be performed in a simple and robust way, amenable for real-life applications.

Future work on such a dual-robot system could focus on improving field-readiness and reliability. A structural revision of the pod, currently manufactured from vacuum-formed shells and 3D-printed parts, would permit faster assembly, more reliable waterproofing and increased compactness. The geometry of the pod could also be optimised for improved hydrodynamics and stability. Additionally, further work could enable the system to become fully autonomous. This could include implementing a position controller for the horizontal plane and a vision-based navigation component allowing more elaborate underwater mission execution. Some form of stabilisation in response to water currents would also be beneficial.

Chapter 13
Practical Tips for Building Aerial-Aquatic Robots

This book would not be complete without a chapter on practical hardware and software elements used throughout the presented robots. We hope that this can serve as a rough toolbox for aerial-aquatic vehicle development, and cover some of the prototyping choices that are often under-reported in academic literature, but consume outsize research time.

This chapter covers several points. The first part focuses on actuation, which is a always challenging to integrate in aquatic environments. The second part briefly presents the composite methods employed in most prototypes. This is particularly important as carbon fibre composites currently offer the best strength-to-weight ratio of any easily available material. In addition, curing carbon-fibre can yield complex geometries and importantly user-selected reinforcement where the highest stresses occur. The third part gives insights into the launching apparatus widely used in experiments, from dive tests to free flight tests. The fourth part covers sensing options for lightweight aerial-aquatic devices, which ultimately will decide the utility of a given system.

13.1 High-Speed Actuation

A significant but often unsung driver of performance improvement in robotics has been the steady improvement in permanent magnet electric motors, driven by improvements in the manufacture of magnetic materials, reduction in the weight of structural components and the refinement of drive electronics. Power densities of 6 kW/kg are now commonplace in brushless motors designed for hobbyist UAVs, and high end electric drive trains now have greater power density than jet engines (gas turbines typically have around 10 kW/kg).

Affordable control electronics allow very high startup torques and precise control of rotation speed, and the improvement in motor control fidelity is a large part of what

R. Zufferey et al., *Between Sea and Sky: Aerial Aquatic Locomotion in Miniature Robots*, Biosystems & Biorobotics 29, https://doi.org/10.1007/978-3-030-89575-4_13

has made multirotor flight possible. In addition, electrical commutation (as opposed to the mechanical commutation found in brushed motors), has made service life much longer, and allows rotation speeds in excess of 10,000 rpm to be sustained long term. The principal advantage of brushed motors is their price, which is significantly lower. But for the design of propulsion systems, brushless motors should be used over brushed motors. For the specific case of aerial aquatic robots, some care is needed where the vehicle geometry prevents air/water flow over the motor, as modern low mass motors use less material and have low thermal inertia, making overheating more likely.

13.2 High-Torque Actuation

By and large, DC motors are best suited to high rotation speeds (a motor is at its most powerful at 50% of its maximum speed, and at its most efficient at around 70% of max rpm), and an application requiring high torque and precision in a small volume requires a gearbox and additional motor control.

Fortunately, the ubiquitous ‘servomotor’ form provides this in a simple package. For servos used to actuate aerodynamic control surfaces, both speed and force are important to the overall control authority, and so peak power output should used as a performance measure, which can be calculated from manufacturer’s data on speed and stall torque (Fig. 13.1A). For the AquaMAV (Chap. 7), the most power dense servo at the appropriate size (4 g) was selected. To actuate larger mechanisms, such as folding wings, speed is often less important than the need to produce sufficient torque to overcome resistance from mechanism friction or, for example, wing drag. To choose such servos, torque per unit weight can be used as an objective, and the most torque dense servo selected (Fig. 13.1B).

13.2.1 The Morphing Case

Most aerial-aquatic robots significantly alter their geometry to adapt to the different media density in their various operating modes. This morphological change inevitably incurs moving large surfaces in challenging condition. Such a movement requires specific actuation with two requirement. First, the actuation requires high torque and low speeds, falling into the category described in the previous paragraph. Secondly, the actuation occurs sparsely, maybe not even more than a few times in the whole mission. Due to the high power limitation of flying vehicles, the actuator mechanism cannot draw power outside of the transition states.

In summary, the following requirements should be met: 1. High torque (speed irrelevant) 2. Fixed locking position 3. Adequate waterproofing 4. Minimum weight.

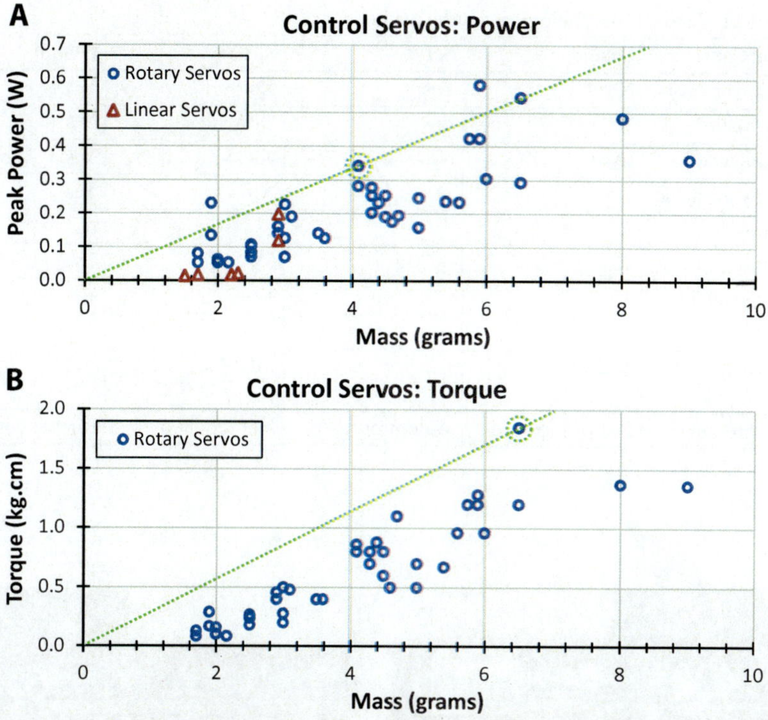

Fig. 13.1 Servo selection example for the AquaMAV (Chap. 7): Key performance measures can be calculated from an assembled database of miniature servos. (**A**) Peak power output against servo mass, with the servo selected to actuate the tail control surfaces marked with a dashed green line. Only two other servos have a higher power per unit mass. (**B**) Maximum torque against servo mass, with the servo selected to actuate the wing folding marked with a dashed green line. The selected servo has both the highest absolute torque and highest torque density

Table 13.1 presents four methods that fit the small size constraints.[1] Weight, as for every component of a flying vessel, is to be minimized. While a Shape Memory Alloy (SMA)-based solution is potentially the lightest solution, and this without waterproofing issues, it comes with many drawbacks highlighted in the table such as a low reliability. Indeed, electrical as well as mechanical contact with an SMA degrades due to the deformation induced in the wire. In addition, performance is dependant on ambient temperature and wind, increasing the difficulty to obtain a repeatable motion. Standard rotary servomotors is by far the simplest solution, usable without any modifications. However, waterproofing comes at the price of weight, and commercial servomotors typically work at a relatively high speed (60deg/0.1s). For structural morphing, this is often not required, favouring systems with higher reduction. Additionally, servo motors are back-drivable, which will induce power usage to keep the position. The lead-screw solution is a viable tested option. Substantial development

[1] 6 cm hull width.

Table 13.1 Studied Morphology actuation methods

Requirement	Lead Screw	Servo	SMA	Folding
Position control	No	Yes	No	Yes
Angle 0–90°	Yes	Yes	Difficult	Yes
Locking at ends	(No)	No	No	Yes
Readily available	No	Yes	No	Yes
Features				
Weight	25 g	> 20 g	5 g	18 g
Torque/weight	+	–	+++	+
Speed	–	++	–	–
Electronics to be added	MOSFET	–	MOSFET	–
Waterproofing	2 slip-rings	Waterproof	Not required	Rubber cone

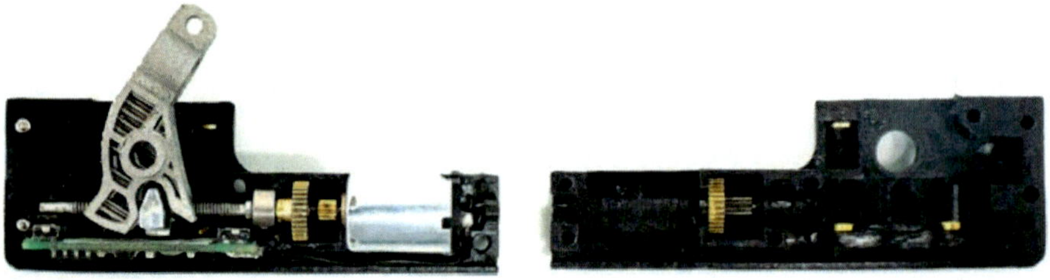

Fig. 13.2 Open view of the Hobbyking folding actuator showing the double locking engineered cam, the control board and the metal transmission

is needed to fully waterproof and actuate a wing without a large transmission weight. Stiffness tends to be an issue, increased at smaller scale.

Micro landing gear servo-actuator are a robust solution for changing robot morphology, featuring 2-position lock and compactness. Such an actuator is shown in Fig. 13.2 and features a rotating stainless steel, laser-sintered *cam*. This pivoting part is locked from rotating in both directions when it sits at either extremity. The cam can be customised which gives freedom of choice on the exact angles as well as cable pass-through and carbon-fibre shaft fixation. A double switch and controller ensures that the lead screw moves the cam from one position to the other. This folding servo-actuator only draws current when rotating the cam, the lead screw provides a high level of reduction and takes 0.7 s to move from one position to the other, see Table 13.2.

Table 13.2 Wing/Sail folding actuation

	Weight (g)	Travel speed (s)	Operating angle	Length (mm)	Cam width (mm)	Voltage (max) (V)
Hobbyking retract	18	0.7	0–90°	55	8	6
Dynam 1149 retract	30	0.7	0–75/90°	70	12	6

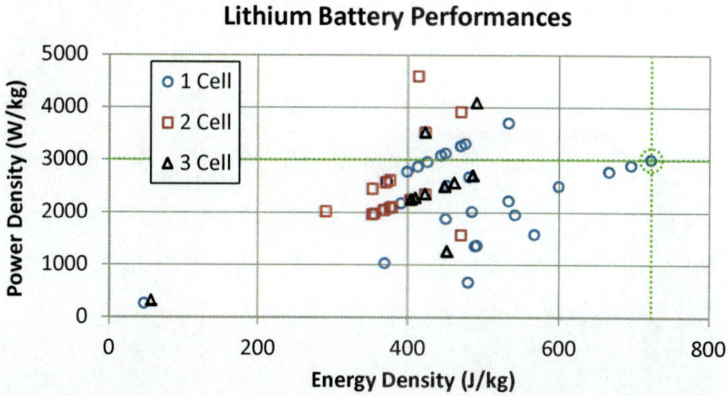

Fig. 13.3 Battery selection: Graph shows energy and power densities of many miniature batteries, and the selected 650 mAh battery is marked with a green dashed line. This battery is the most energy dense and has a maximum current of 10 A

13.3 Battery

When selecting a battery, there is typically a minimum size that can be considered, defined by the peak current requirements (e.g. the 9A needed to actuate the SMA wire in Sect. 6.1). Beyond that constraint, the performance of many different batteries of appropriate should be gauged in terms of total stored energy per unit mass, and maximum output power per unit mass (Fig. 13.3). The most energy dense battery would improve the vehicle range, and in the case of the AquaMAV, a 650 mAh single cell battery was selected and two of these cells were used in series to produce sufficient voltage to drive the electronics.

13.4 Ignition

Ignition of combustion engines has become an extremely reliable process, running for millions of cycles without the need for maintenance. However, there are key differences when attempting to trigger a single reaction in a new prototype such as the consecutive jump-glider (see Sect. 9.3). Firstly, the gas mix is at atmospheric pressure

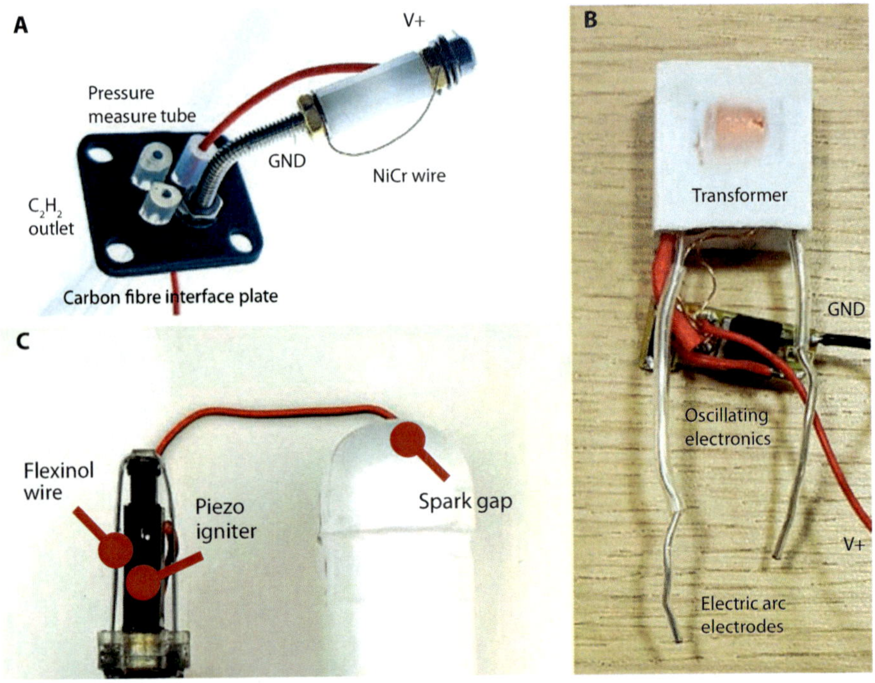

Fig. 13.4 **A** Glow Wire ignition system mounted on the robot's interface plate. **B** High voltage arc generator before robot integration. **C** Piezoelectric ignition system using flexinol

which makes ignition harder than at the higher pressures achieved in petrol engines. Secondly, the chamber is at ambient temperature also reducing flammability. Lastly, the main factors hindering ignition are a high level of humidity and the presence of water. This in particular has been identified to cause problems of splashing on the ignition system, which can be problematic. All those elements drastically increase the complexity of reliable ignition. Three methods were identified, designed and tested as working solutions to igniting a acetylene gas mix.

Piezoelectric sparker A piezoelectric sparker generates an electric arc from a high force, small motion mechanical actuation. They are commonly found in lighters and are usually triggered by hand. However, they can also be actuated by the means of a suitable electromechanical device. A shape memory alloy works well for this situation, giving a few percent elongation at a high force when heated through Joule effect. This method is robust and reliably produces a spark. However, the SMA wire requires a high current to operate and the attachment points break over time. In addition, no feedback mechanism indicates whether a spark has been generated, and as such, precise timing is difficult.

Electric arc An electric arc is generated when the voltage exceeds the breakdown threshold given by:

$$V_{req} = \frac{a p_a d_{gap}}{\log(p_a d) + b}$$

Parameters a and b for Paschen's law for air at STP are 4:36e7 respectively 362e3. d_{gap} is the distance between the two electrodes, which has a lower bound imposed by the chemistry of the gas, calculated in the following relation [204].

$$d_{gap,min} = \left(\frac{6 * MIE}{\rho c_p \pi \left(T_{product} - T_{initial} \right)} \right)^{\frac{1}{3}}$$

With MIE the minimum inition energy and ρ the density of the gas. In the case of acetylene, a minimum arc distance $d_{gap,min}$ of 0.4 mm is required. This implies a minimum voltage of 3.5 kV.

One noticeable issue the arises with this pulsating arc system is interferences with the wireless communication system. This manifests as a reset of the wireless chipset, forcing the end-user to reconnect and losing connection for a moment. This issue is mitigated by the addition of a copper shield around the transformer and high voltage wires. The plasma arc itself is positioned as far as possible from the rest of the electronics.

Sparking "barbecue lighter" style systems are widely available and reliably produce a single spark. However, they employ a large multi-stage transformer that prohibits its use in a lightweight vehicle.

Glow wire An alternative method for igniting a gas is via a hot element. The ignition temperature of acetylene is 305 °C, which is achievable by a standard NiCr wire (melting point of 1400 °C). The wire needs to be clamped instead of soldered, held by a heat resistant structure. Ignition is reliable if the wire reaches a high temperature, emitting a bright red glow. However, the wire degrades over the cycles which changes its resistivity ρ and hence the power emitted (13.1). This in turns reduces the temperature of the wire.

$$P_{wire} = \frac{V_{bat} A}{\rho L} \tag{13.1}$$

A measured 1.5 A of current was required for ignition, varying with wire length. Precise timing is difficult due to the changing conditions of the wire and the environment. A light breeze on the wire already greatly increases convection cooling and therefore vastly changes the temperature evolution. Experiments have shown that the wire is susceptible to breaking from the combustion process, potentially linked to the pressure front propagation and thermal shock. Strong acetylene combustion events breaks the wire after 2–5 cycles. This reduces the usability of such a mechanism for a long-lasting device. However, it is worth noting that the wire is straightforward to replace, and as such is an adequate simple solution for laboratory single tests.

13.5 Waterproof Openings

Waterproofing is challenging in the most favourable conditions, and the presence of buoyancy makes the mass budgets of aquatic systems far more forgiving than those of flying vehicles. The thick, highly stiff structures of conventional underwater hulls can be reliably sealed against water ingress, but the lighter structures need for flight are inevitably more flexible, which undermines the efficacy of clamped rubber seals. This is particularly true of airframes impacting the water, which experience very high loading that is likely to flex seals and allow water penetration.

Due to the problems of reliable sealing, judicious design is useful in slowing any potential leaks. This can be achieved by having only a single possible leak point - multiple sealed hatches will allow simultaneous escape of air and entry of water, which accelerates water penetration. Care should also be taken to keep seals away from areas of high water pressure or structural stress.

When salt water is present inside the hull of a robot, even in small quantities, it will rapidly destroy off the shelf electronics, with corrosion progressing more quickly due to galvanic effects and the presence of a battery. Non-conductive cleaning fluid should accompany marine field work, so that robots can be stored without salt. 'Potting' or coating electronics is effective protection, but care must be taken with thermal management, as most flight propulsion electronics rely upon airflow for cooling, and additional heat sinks may be necessary.

The most challenging component to seal is the motor. The labyrinth seals commonly used in submarines and boats (commonly called 'stuffing boxes') prevent water ingress around rotating shaft bearings by making the water entry pathway too long and narrow for the water to penetrate. However, labyrinth seals are typically large and heavy, and often add too much weight to the drivetrain for an aerial-aquatic system. Magnetic couplings are very effective at maintaining watertight seals, but are again heavy. For small scale robotics, the best approach is often to simply leave the motor in the water, and run power through a watertight cable gland. Most brushless motors are intrinsically waterproof (motor wires are coated in shellac and housings are made from anodised aluminium). Care must be taken with corrosion, in particular the magnets and shaft bearings. The former can be coated and the latter can be removed and replaced with corrosion tolerant components if necessary.

13.6 Composite Construction for Aerial-Aquatic Robots

Aerial-aquatic robots are severely limited in terms of payload due to their need to fly. As consequence prototyping these robots benefits tremendously from carbon fibre composite materials, which effectively provide the strength of steel with the density of a polymer. Indeed, these materials possess the highest strength to weight ratio commercially available and have therefore become ubiquitous in high-end prototyping.

Table 13.3 Comparison of thin-layer carbon-fibre manufacturing techniques

Components	Minimum thickness (μm)	Manufacturing time
Pre-preg	>200	Low
Woven layup	100	Medium
Unidirectional layup	20	High

They are typically well-suited to research, where single prototypes are manufactured and labour per unit is of less consequence.

Composite structures consist of a fibre matrix bonded to a resin. The fibre matrix can be arranged in many ways, such as random short strands, woven fabrics and unidirectional fibre layouts. The resin play an important role, giving a strong rigid product or also a soft bendable composite. Carbon-fibre products can be either obtained in cured, finished form or as separate fibre and resin. Cured products come in various shapes, typically plates, rods and tubes which can be further cut, drilled and otherwise adapted. For example, 2D cutting of carbon fibre plates is common practise, as exemplified by most high-end drone structures.

In the case of more complex shapes, carbon fibre is cured in a mould. This is where composites truly shine as it gives immense flexibility to the engineer. The fibre can be arranged in the orientation that will better take the structure load. Local reinforcement is also one of the benefits.

Several techniques exist to manufacture composite parts (Table 13.3) but perhaps the most straightforward technique for small scale prototyping is hand laminating (wet-layup). It consists in applying manually one or several layers of fabric impregnated with resin onto a mould, giving the user choice over fabric as well as resin. One can also use Pre-impregnated fabric (pre-preg); this way no resin application is needed on dry fabric. In the following subsections, we quickly introduce those different methods.

13.6.1 Prepreg

Pre-preg is the fastest method for obtaining an excellent cured carbon-fibre part as the resin (pre-cured) is already contained in the right proportion and matched with the fabric which guarantees uniform result (optimised by the manufacturer). However, minimum thickness is limited and the thinnest commercially available pre-preg is of 0.2 mm thickness.[2]

Woven fabric is shows high rigidity in all directions as opposed to a unidirectional fabric as the fibres are oriented in several directions instead of one. Therefore, to obtain good stiffness properties, several unidirectional layer of fabric need to be stacked. However, this requires more manual work compared to hand laying one layer of woven fabric.

[2] Easycomposites Pre-preg X110, 210gsm.

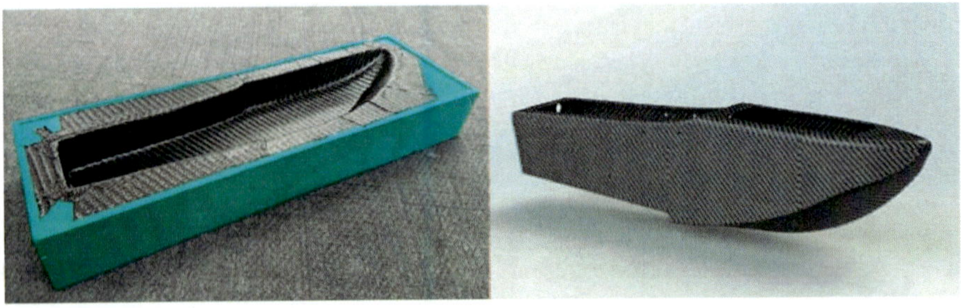

Fig. 13.5 Mould manufacturing of the hulls

Waterproof hulls fabrication The hulls of the sailMAV presented in Chap. 11 are the most critical part of the whole robot as its dimensions play a central role in mechanically linking most other components together. Although a composite hull is a lightweight and robust option, the concave shape severely reduces the ease of manufacturing. Most problematic are the top edges of the hulls, that partially close the shape and makes it impossible to employ a rigid inside mould. To overcome demoulding difficulties, the hulls are length-wise split in two halves (Fig. 13.5 left). Outside moulds are used as they yield an excellent finish in the outer surface, where the water contacts the structure. The moulds are CNC-cut from a block of dense epoxy and the surface sanded, sealed, polished and coated with chemical release agent. Pre-impregnated fabric can be easily sourced in thicknesses of a minimum of 210 µm which provides sufficient strength as shown in previous robot developments. The fabric is laid-up in the mould following an inside-to-outside workflow (Fig. 13.5 left). The bottom surface of the mould consists of a continuous patch of pre-preg while the walls are rectangular patches overlapping the bottom patch by about 4 mm. This gives some freedom of movement in the sharp bottom edge, i.e. the overlapping fabric can stretch in those difficult zones. The two halves are sealed inside of a vacuum bag and cured in the oven.

The two halves of cured composite are bonded together (Fig. 13.5 right) using epoxy and a strip of fabric is cured in the inside, over the bond line to guarantee structural integrity.

High pressure combustion chamber fabrication On the Aquatic Jump-glider (see Sect. 9.3), the combustion chamber is the most critical mechanical element, needing to withstand up to 10 bars of pressure. The chamber is a thin-walled pressure vessel the hoop stress in the wall can be approximated by the Young-Laplace equation:

$$\sigma \approx \frac{Pr}{t}$$

where σ is the hoop stress, P is the internal chamber pressure, r is the chamber radius and t is the wall thickness. Using this equation, with a maximum chamber pressure of 750 kPa, and a tensile strength of the carbon fiber composite of 645 MPa, the required thickness can be calculated to be 35 µm which can be achieved with a single layer of prepreg woven carbon fiber composite, yielding a minimum thickness of 200 µm. This allows a considerable margin for manufacturing defects, as well as stress concentrations from the chamber design, which includes two flat sections to hold the electronics capsule, visible in the top part of Fig. 9.10B, and the wing.

Carbon fibre pre-impregnated with resin is wrapped around the two moulds. Special care is taken that the necessary cuts in the fibre do not overlap. Once the top and bottom shells are cured and removed from the mould and the overlap section sanded. Finally, the shells are bonded together.

13.6.2 Woven Wet Lay-Up for Wings

The wings built for the AquaMAV prototype (Chap. 6), were cambered flat plates, chosen to avoid the high buoyancy created by thick wings. These were produced with a simple sheet of aluminium, manually rolled into an airfoil section with a slip roll, which allow moulds to be produced quickly and cheaply. The composite itself used pre-preg sheets, and an entire wing could be made with less than an hour of labor. However, this simple construction technique resulted in structurally poorly-performing wings, owing to the lack of torsional stiffness of the cambered sections. Ultimately, for later aquatic aircraft, a wet layup approach was adopted, so that stronger, more stable wings couold be produced.

The lightweight wings of the SailMAV UAV (Chap. 11) are built around cores made out of foam[3] with a low density of 31 kg/m^3. The foam cores stay in the wing and allow to have a full and rigid structure. The foam cores were cut using a CNC router. This technique allows to obtain a very accurate airfoil shape and a smooth surface, unlike the commonly used hot wire technique with which it is very difficult to obtain a precise results. Manufactured from two halves, the wing is bonded together before wrapping.

Wet Layup consists in applying resin manually which is already mixed with the hardener onto the dry fabric. This method yields good result with high strength-to-weight if the following steps are respected, which are based on discussion with professionals and experience.

The wet layup technique showcased in Fig. 13.7 is an effective method for thin-fibre fast prototyping, requiring about 1–2 h of manual work for a regular wing. The best result is obtained with moulds that are machine-cut (CNC) which can be a lenghty process. Extremely thin fiber can be selected, down to a 0.1 mm woven dry fabric. While thinner fabric does exist as unidirectional fibre, a woven cloth provides the best strength in all direction and reduces delamination. Faster manufacturing

[3] IG-F Rohacell.

(a) No mould (b) Metal sheet mould (c) Epoxy mould

Fig. 13.6 Mould impact on composite wing structure

Fig. 13.7 Key stages of a wet lay-up process. **(left)**. Machined foam core on top of the bottom mould. **(middle)**. Foam core is applied to the resin-wet fibre. **(right)**. Side-view of the mould stack before curing under pressure (vacuum)

methods were tested without 3D molds, instead relying on a bent metal sheet or even nothing at all (see Fig. 13.6). Those methods could work on stronger foam or when a smooth surface is not required.

A central foam core with high strength, low density is required since it will completely fill the wing. A low-temperature cure resin should be selected because the foam core will not sustain high temperatures. An excess of mixed resin is poured onto the flat-laying fabric (roughly cut to size). The resin is squeezed out on the sides with a flat, blunt tool until the resin is properly wet yet no extra resin is present. Then wrapped around the foam core, it is carefully pushed into the bottom mould. The top side of the mould is closed, taking special care that the wet fabric is stretched to not pinch the leading edge. The whole assembly is cured under pressure which can be applied by (a) a vacuum bag in an oven or (b) a hot press. Cure needs to happen according to the temperature profile indicated by the manufacturer.

13.7 Launcher Testing Apparatus

Aerial-aquatic robots need to function in both water and air as well as perform the transition between media. A new tool is required to better study the questions that arise during the aerial phase and the air-to-water transition. Particularly winged flight requires the robot to achieve a repeatable velocity. Featuring a 150 cm-long rail, it is designed to accelerate a robot of up to 1 kg to speeds up to 20 m/s. This research tool has been used extensively in a multitude of situations to fly unpowered robots, study the impact of objects into the water as well as slowly move camera systems.

Working principle The launcher consists of a linear rail with a carriage sliding on a linear bearing seen in Fig. 13.9 (left). The robot or accelerated object is attached to a carriage (in blue) via application specific holders. The whole rail structure can be tilted from 0° to 90°, shown in the illustration at 60°. This range not only permits flight launches at all angles but can also accelerate robots downwards to study their impact into water. The structure is mounted to a wooden triangular frame with wheels that can be easily moved.

Transmission A transmission belt accelerates the carriage to set terminal velocity by the end of the rail. The steel reinforced toothed belt spans between the two wheels afixed to the extremities of the rail. The bottom wheel has 48 teeth and is driven by an electric motor. During the acceleration phase, the clutch is engaged (Fig. 13.8 position 1) and the belt pulls the carriage without slipping. As soon as the side tab impacts the clutch plate (Fig. 13.8 position 2), the central cam lifts up and released the downward pressure that was previously keeping the belt engaged. The front and back lobes guarantee that the belt cannot re-engage randomly. The carriage now slides freely (Fig. 13.8 position 3), disjointed from the transmission, and hits the shock absorbers. The carriage passes in front of a light gate which sends the shut down signal. In less than a second, the transmission is electrically stopped and the energy dissipated in the power unit, instead of the shock absorbers.

Figure 13.9 (right) shows the clutch mechanism designed for this launch system. Machine cut from a single block of aluminium and anodized for improved surface hardness, the carriage screws into the rail slider. A central cutout is created to house the pinching cam. Due to an offset rotation axis, it allows a pinching travel of 3 mm, equivalent to the tooth height. The stainless steel cam is driven by the two tabs fixed

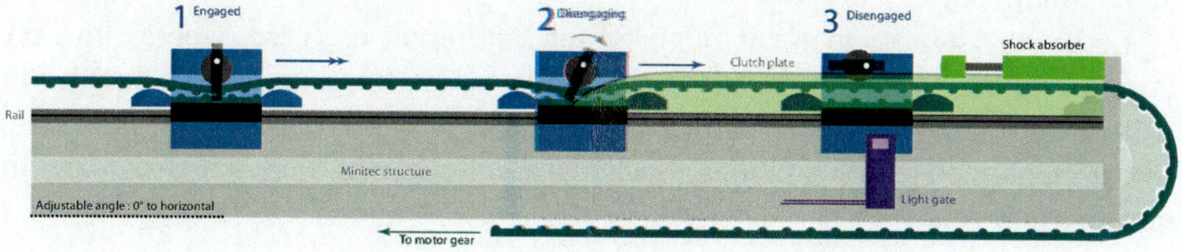

Fig. 13.8 Working principle of the transmission mechanism of the launcher system

Table 13.4 Tested electric motor configurations

	Induction AC	Brushless	Stepper
Use case	Reliable, low-powered launch, does not require custom electronics	High-speed launch on battery power with high level of control	Precise camera movement, optical flow testing, linear object motion
Advantage	Reliable, industrial, pre-configured	High-power, flexible control, battery powered, wireless control	Precise timing and speed control. Very slow speed and high torques
Disadvantage	Heavy, driver module not designed for fast stopping. Feedback loop fixed at a 100 Hz. High rotor inertia	Unreliable at low speeds. Additional micro controller required	Inefficient, low speeds do not permit launching
Motor	500 W industrial AC	4000 W BLDC	2A large stepper
Driver	AC inverter	VESC 4.12	T6600 H-bridge
Control	Wired Remote with switches	Micro-controller with Bluetooth	LCD interface with Arduino
Speed	50 Hz max	150 Hz max (no load)	Slow 5 Hz
Voltage	230 V	48 V	20 V
Price	800 $	300 $	150 $

to the extremities of the shaft, visible on the outside of the carriage. The carriage is designed to be mounted without removing the belt due to gaps next to the lobes. Finally, the lower side plate serves to block the optical sensor at the end of the trajectory.

The launcher mechanism can be operated on various motor configuration depending on the use scenario. Table 13.4 lists the differences between the an AC induction motor, a brushless DC motor and a stepper motor.

The brushless outrunner motor relies on an Electronic Speed Controller that switches the phases. The outrunning rotor carries the permanent magnets. Very lightweight, this solution is affordable and has low rotor inertia which makes it ideal for high accelerations. The ESC is controller by a separate uC unit with wireless communication. As such, instructions are sent from a phone. The light gates at the extremities of the rail trigger interrupt signals in the uC that decelerates the system at maximum power.

Lastly, the stepper motor is intended for a different use case, where slow, precise rail motion is required. This is useful for camera motion, optical flow testing and various situation where known velocity and position is important. All of the above-mentioned motor solutions attach to the main driving shaft which actuates the transmission belt.

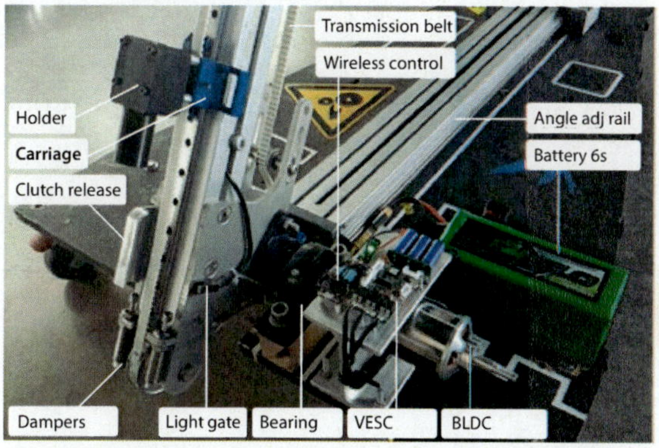

Fig. 13.9 (left) Overview of the launcher assembly. (right) mechanical linear anodized aluminium clutch assembly

Fig. 13.10 Launcher flight composition

Flight tests with the launcher The launcher has been at use in many scenarios, such as initial velocity flights for the Aquatic Jump-glider. Replacing the chemical, high-pressure thruster with an equivalent inert thruster significantly reduces the complexity of a launch. Instead, analysis and understanding can focus on the jumping performance that follows the initial acceleration. A large amount of launches were effectuated at varying velocity, angle and wind conditions. Figure 13.10 shows a composition of the robot's trajectory with light tail wind at a 10 m/s initial velocity.

13.8 Sensing Options for MAVs

Aerial-aquatic robots have the potential to become field companions that can autonomously collect samples and data for scientists. This environment monitoring task is already identified and described in Sect. 2.2. One of the challenges to

Table 13.5 Water environment sensing options

Measure	Application example	Output	Sensor
Water temperature	Calibrate satellite data, relevant for climate models	°C	TSYS01, MAX30208
Salinity	Iceberg vicinity measure	%	Conductivity sensor
Turbidity	Measure of sediment density	%	Optical sensor
Oxygen	Measure of fluxes in and out of the ocean for climate models	O2 concentration	Optical sensor
Chlorophyll	Plankton blooming evaluation		Fluorometer
DNA sequencing	Biological applications	Water sample	Micro-pump
Underwater imaging	Wildlife, reef imaging	Image	Smartphone camera
Aerial imaging	Imaging and tracking of large animals in water, comparison with satellite imaging	Image	Smartphone camera
Atmospheric	Relevant for comparison with water temperature	°C, %, Pa	BME280
Hydrophone	Listen to the tags of animals in water	Db	
Infrared	Compare the IR signal from water and air for satellite calibration	Spectrum	IR sensor 2x

complete this mission profile is the integration of sensing packages that are both (a) relevant to the final user and (b) does not impede flight and transitions.

There is large gap to be bridged between current marine sensing technology and lightweight aerial robots. Most marine sensors are heavy, pressure-resistant and rugged by design. Yet, this is highly incompatible with the gram to kilogram scale payload levels available on aerial-aquatic MAVs. As a result, direct integration of commercial marine sensors is almost always impossible. Instead, hardware sensing solutions can be found in drone technology, smartphone technology or in-lab developed miniaturised technologies.

Through joint discussions with environmental scientists and sensor-oriented researchers, we have synthesised a list of sensing packages. Table 13.5 identifies sensor options that could be suitable for lightweight aerial-aquatic platforms. A few of these sensors have been already implemented in the robots presented in previous chapters with promising results.

Chapter 14
Conclusion

14.1 Summary

This book introduces the concept of small, unmanned aerial-aquatic robotics. This novel field of research aims to merge the benefits of flight and aquatic operation into one lightweight autonomous platform. As the reader will have seen in this book, wildly different robots can be envisioned as solutions to this formidable challenge.

To start with, we presented in detail the motivation of studying this topic, providing examples of specific practical applications of the technology but also of the current research questions in the field. In particular, the coexistence of aquatic and aerial features in one robot introduces difficult design, operation and control challenges. The large difference in density between water and air, in particular, is a key factor at the root of this difficulty, making the transition across the interface an exciting research question in propulsion, design, autonomous control and other topics.

The transition from water to air and vice-versa is certainly the most challenging step towards hybridising unmanned systems. Therefore, this book has placed a strong emphasis on this aspect and we have discussed three different methods to perform such manoeuvres: **Impulsive**, **Gradual** and **VTOL**. Impulsive transition employs an addition power source to quickly reach flight height and speed. This additional power source can be either pre-loaded, such as with the CO_2 compressed gas on the AquaMAV or generated on board, such as in the Aquatic Jump-glider. Gradual transitions refers to the acceleration of the robot to sufficient speeds to be fully airborne. Practically, this can be done through hydroplaning at the water surface, through secondary hydrofoils or even through streamlined geometries that permit high speeds underwater, as observable in jumping fish or flying squids. Finally, Vertical Take-Off and Landing vehicles can transition from an underwater hovering state to an aerial hovering state by increasing their propeller thrust momentarily. Air-to-water transitions can somewhat analogously be classified as **gradual horizontal**, i.e. landing and decelerating on the water surface, **gradual vertical**, i.e. landing on the water surface at very low velocity, then sinking into the water, and **impulsive**, i.e. diving into the water at high speeds, e.g. through a plunge-diving manoeuvre.

R. Zufferey et al., *Between Sea and Sky: Aerial Aquatic Locomotion in Miniature Robots*, Biosystems & Biorobotics 29, https://doi.org/10.1007/978-3-030-89575-4_14

Table 14.1 Operation envelope of the prototypes presented in this book

Aerial	Aquatic	Transition	Demonstrator
Winged flight	3m passive dive	Impulsive, single	AquaMAV
Winged flight	Winged	Impulsive, single	AquaMAV w/gear
Glide	Surface	Impulsive, multiple	Aquatic jump-glider
Winged flight	Surface + sailing	Gradual, multiple	SailMAV
Multirotor flight	5m + 3D motion	VTOL, multiple	Medusa dual-robot

Table 14.1 summarises the different technical solutions discussed in this book with their transition capabilities. The robot examples discussed in this book for each case are also listed.

The five solutions detailed in this book feature very different operation modes. The aquatic operation envelope ranges from passive reloading at the water surface to full three degree-of-freedom motion for the pod of the MEDUSA dual-robot. Extended capabilities always come at the cost of increased weight and complexity. Ultimately, the specific application determines which locomotion scheme is the most adequate.

While initial research has been undertaken on autonomy and control of small-scale aerial-aquatic robots, most work to date has focused on the design aspect, addressing the challenge of obtaining a vehicle physically capable of aerial-aquatic operation. Building on the considerable progress achieved in the design domain, future work is expected to also venture further into the control domain, with the ultimate goal of enabling reliable autonomous mission execution. This will require increasingly reliable platforms and designs, continued advances in hardware, and novel control and sensing approaches. Dealing with uncertain and challenging environmental conditions will pose a particular challenge and will be essential to address as the vast majority of missions envisaged for aerial-aquatic vehicles take place outdoors.

Figure 14.1 qualitatively sums up the current level of control and autonomy of the discussed aerial-aquatic robots as a function of scale. Unsurprisingly, larger robots feature higher levels of autonomy and control, at the expense of cost, safety and ease of deployment. The payload of a given robot scales with its mass and as such, more electronics and sensors can be fitted on a heavier robot. This also implies more computational capacity, which in turn permits more advanced control. Nonetheless, electronics have seen an extraordinary amount of miniaturisation in recent years, in no small part thanks to the rapid development of smartphones. The plethora of smart devices that are continually emerging feature an increasing amount of sensors, more accurate and condensed than ever before. These technology advances have made flight controllers, micro-processors, flight sensors, and Inertial Measurement Unit (IMU) devices extremely lightweight, and the trend is likely to continue. The high-level trends, features and limitations observed on the robots in this book can be generalised to other similar platforms, and thus provide a guideline for future developments in the field.

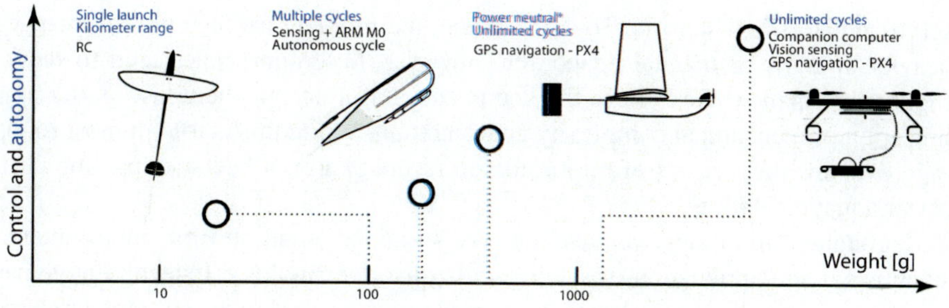

Fig. 14.1 Control and autonomy of different robots. *power recovery has not yet been experimentally demonstrated on the SailMAV

At 160 g, the AquaMAV robot is to date one of the lightest of the aerial-aquatic family. In order to keep the mass to a minimum yet employ commercially available parts, the robot's electronics are limited. The AquaMAV carries all the required circuitry to fly remotely, giving it full power autonomy. Due to the lack of a microcontroller onboard, the robot cannot fly autonomously. With the advances of available lightweight electronics, a robot of this scale will likely soon be able to fly autonomously.

The Aquatic Jump-glider features a microcontrolling unit that operates the robot with basic autonomy. This robot follows a launch-flight sequence that is based on timings while simultaneously recording flight data. However, the flight itself is not auto-piloted with localisation.

Increasing the scale of the robot further with payloads of 10's of grams enables the integration of all the required sensors and computation for full autonomy. The SailMAV vehicle, at around 700 g, can hope to achieve this through tight integration with PCBs.

The MEDUSA platform, with its large kilogram-scale payload, can afford to carry both a flight autopilot and a dedicated companion computer. This enables faster developments, more advanced levels of planning, autonomy and cross-platform reusability.

14.2 Lessons Learned

During the years of research that this book presents, we have learned several lessons (sometimes the hard way) about aerial-aquatic robots and their development. We hope that the following points will enlighten the reader about issues, mistakes and wrong beliefs that we have encountered.

A simple yet considerably underestimated difficulty that we were constantly faced with is the incompatibility of small flying robotic systems with water: an aspect that bridges various levels. Firstly and more importantly, most hobby-grade components widely used in drones are not waterproof. This is an issue for example with servo-actuators and sensors. To some extent, the problem can be circumvented via heavier systems or manual adaptation, however these approaches tend to reduce flight performance. We believe the key to successful designs here lays in the minimisation of mechanical complexity and actuation. While this is true in most robots (and explains the success of multirotors in recent years), it becomes paramount in aerial-aquatic solutions.

Throughout the years, we have received sustained attention from environmental scientists in the application of small aerial-aquatic technology. Indeed, a large part of aquatic data collection is based on manual processes and relies on often large and outdated tools. Small, affordable, autonomous and quick aerial systems could vastly enhance data collection. While the interest of other scientists has certainly motivated us to improve our robots, it has also shown us that their required level of reliability is high. The Technology Readiness Level (TRL) of the field would need to be further increased to perform missions that can benefit other users. This hinges on a variety of points, namely the autonomy level, the resistance to perturbations, communication levels with the user, sensing package etc...

Quick design-test-evaluate iterations help accelerate research in mobile robotics. This paradigm has been successfully applied to many sub-fields of robotics and aerial-aquatic robotics is not an exception. However, testing such systems is rendered more difficult by the necessity of proper sealing against water ingress and wireless operation. In our experience, testing occurred far less often than we would have liked and every cycle implied lengthy experimental preparation. It would perhaps be advisable to extensively validate sub-systems before attempting complex maneuvers. This has been done with the AquaMAV where a propulsion-only prototype was first tested before adding flight capability.

The difficulty to undergo quick iterations has been amplified by the lack of available space for experimentation. Very few places in the UK and Europe permit drone flights above water bodies. Special authorisation is often required, generating additional administration overhead. Particularly testing autonomous vehicles and missions of longer range and duration can be challenging.

Last but not least, aerial-aquatic robotics is a field where mechanical failure tends to lead to disastrous consequences for the whole robot, further increasing the difficulty of experimentation. Indeed, it is hard to stress how much damage occurs upon unexpected submersion of a flying robot without adequate protection. The same can be said about the consequences of a forgotten seal or a broken screw in the robot body (fuselage). Unsuccessful or inaccurate manoeuvres can have dire consequences, as for instance in the case of a water entry diving manoeuvre terminating on the beach instead of the water surface. Indeed, the combination of unconventional designs, unconventional manoeuvres and multi-modal operational environments implies that even relatively minor shortcomings can cause irreparable damage.

To conclude, we would like to share two important lessons based on our experience in the field. Firstly, multi-modal locomotion requires addressing many different physical and engineering challenges, requiring innovations in electronics, mechanics, materials science and other disciplines. To this end, a multi-disciplinary team is a true advantage in designing aerial-aquatic robots. The strength gained from uniting students and experts with diverse backgrounds is key for rapid iteration and successful development. Secondly, the physics of aerial-aquatic robots change greatly in the course of the locomotion cycle. Hence, intuition alone is a poor design driver and representative physical models are an essential tool in the design process. Additionally, the manufacturing of full prototypes is time-consuming, in part due to completeness and proper sealing requirements. As such, the availability of accurate if simple simulation setups can reduce costly design errors.

14.3 How Far Have We Come?

Aerial-aquatic robotics started in earnest with the AquaMAV vehicle, launched in 2012. This concept inspired the multiple subsequent projects described in this book. This body of experimental work represents several research milestones and technical innovations.

Firstly, an analytical model of a pressure-based aquatic escape provided a basis for the development of a working prototype. Secondly, impulsive aquatic escape was experimentally demonstrated using compressed CO_2 as a drive medium. In parallel, the modelling of the transition problem was further studied to better understand the transition across the water surface. Thirdly, the impulsive actuation mechanism was implemented into a full aerial-aquatic vehicle, via adaptive wing morphing. Through the use of water-reactive fuel, the impulsive actuation allowed us to demonstrate consecutive water jump-gliding in 2019. Aerial-aquatic robots were also shown to successfully leverage wing energy to passively move at the water surface for the first time. This led to the demonstration in 2020 of a dual-robot system with a larger flying vehicle can deploy an underwater active probe.

In all the above-mentioned concepts, a common research approach can be extracted. Namely that hybridising locomotion can be approached in separate steps before linking them together. Broadly speaking, the full locomotion cycle can be classified into 4 phases: flight, aquatic, water escape, water entry or landing. The water escape being the most challenging phase, it received the most attention, as reflected by the content distribution of this book. The other stages were subsequently demonstrated before integrated into one full-cycle capable robot. This full cycle needs to be performed outdoors due to the flight space required. This adds a level of uncertainty with the presence of wind, cold temperatures, humidity, waves and currents. Altogether, even after individual stage validation, it still remains a tremendously challenging task to perform a full cycle with current aerial-aquatic technology due to the low reliability of each stage. While the field has evolved quickly in the recent years, significant robotics progress is still required to bring Aerial-Aquatic vehicles closer to autonomous and reliable operation.

14.4 Impact of Aerial-Aquatic Robotics in the World

14.4.1 Environmental Science

The climate is changing, and we need better monitoring technologies to disambiguate the many factors which influence the way in which it changes. Ultimately, better data allows us to focus harm reduction effort where it can best stem biodiversity loss and economic damage. A mobile robotic platform that operates in both air and water would provide an invaluable help in climate understanding, emergency relief and experimental biological research, and a field-ready system is already sought after by environmental scientists.

Aerial-aquatic robots provide the means to cross large distances, obstacles or change from one body of water to another. At the same time, they are capable of operating underwater, collecting samples, video and water data. Combining air and water locomotion in a single robot enables important applications in oceanography and climate research, ranging from calibration of satellite images to microplastic count, fish conservation, coral reef monitoring and sediment counting in arctic rivers. Indeed, current monitoring methods are slow, manual, expensive and often happen in dangerous locations for humans, such as oil spills, the vicinity of icebergs, polluted rivers, floods and contaminated lakes.

Aerial-aquatic robotics have the potential to benefit a large spectrum of marine scientists, professionals or recreational users. Examples include (but are not limited to) conservation biologists, fisheries managers, recreational fishermen, microbiologists, pollution control officers, remote sensing experts, oceanographers and environmental engineers.

Within urban environments, the supply of potable water and tracking degradation of fresh water sources is a constant task that grows more difficult and expensive as the population of cities grows. A flexible, rapidly deployed monitoring tool can provide considerable social and economic value by improving access to water data.

14.4.2 Emergency Response

Aerial-aquatic systems provide unique ease and rapidity in accessing direct water samples than can be achieved with conventional systems, especially in emergencies, where access is limited. To conclude this section, we have outlined three examples of emergency incidents where the unique capabilities of aerial-aquatic systems offer particular benefits to the overall response.

Marine Accident In the event of a stricken vessel contaminating the surrounding water, aquatic aircraft would enable a rapid initial risk assessment, providing immediate spill data so that appropriate wider response can be formulated.

A plausible response mission could include a surface vehicle that enters an area and deploys several aquatic aircraft units to the spill site, which would then enter spill site and return data from surface, with specific chemical sensing tagged with CTD and position data. Estimates can also be made of current from vehicle drift. After initial assessment, operator identifies areas of interest and aerial-aquatic robots travel on the water to ingest relevant surface/subsurface samples and then return to surface vehicle for retrieval.

Water Contamination During Flooding Here, an aquatic aircraft enables acquisition of samples and data from isolated and inaccessible bodies of water to assess potential contamination during a flooding event:

During a flood, a surface vehicle is deployed to begin surveying an area, but during the event that contamination is detected but close survey is impossible due to obstacles in the water or other impassable area (fast flowing region, oil fire, etc.), an aerial-aquatic robot is deployed to traverse the obstruction in the air and return samples and data to the surface vehicle for relay. An aerial-aquatic robot can also be deployed to gain aerial perspective and provide visual information to operator for route planning and hazard assessment.

Chemical Biological Detection The manner in which chemical and biological agents are released has a strong impact on their detectability. A slow release from a moving body results in a diffuse cloud which is difficult to detect with miniature sensors with lower fidelity, and requires high accuracy sensing able to detect low concentrations.

However, dispersal from a single source can result in a drifting, concentrated plume occupying a much smaller area. This is more difficult to detect with a single sensor system, and is best measured either with a distributed set of sensor nodes or a highly mobile sensor platform. Aerial-aquatic robot systems combine both of these modalities into a single system, and can provide both the high vantage point and mobility of flight with the ability to contact the water and provide direct data.

14.5 Technological Trends and Future Work

Improved transition The transition from water to air remains a difficult task for all the presented robots in this book, adding considerable mass and requiring by far the largest portion of development time. While we have presented working solutions, further research is required to develop new robotics methods that (a) are simpler to manufacture (b) are more resilient to unpredictable water surface state (c) have higher transition efficiency and (d) can recover in case of failed transition.

Environmental Energy The SailMAV robot concept illustrates well that wind energy can be used in robots to passively move at the water surface. We are convinced that renewable energies will play an increasingly important role in mobile robotics. This is especially true in aerial-aquatic vehicle, which operate in places where wind, solar and wave energy are available. The rapid progress of photovoltaic cells will trigger further innovation mobile robotics and enable increased locomotion capabilities.

Autonomy & Payload Overall, small flying robots offer only a low payload, which is often incompatible with marine sensors. Indeed, such sensors tend to be heavy and robust to withstand rough ships environment and depth. In order to bring marine sensing technology onto small aerial-aquatic robots, two obvious axis of research are required. Firstly, the payload capacity should be maximised on a given robot. Secondly the sensors themselves should be adapted to an airborne platform, reducing their weight. The applicability of robots is often dependent on high autonomy, both in terms of flight distance and control. Autonomous flight capability is required when flying Beyond Line Of Sight (BLOS), typically occurring around 1 km. Robots operating past this limit need to be equipped with adequate flight controllers and redundancy. The same logic applies to underwater operation and to transitions occurring far from the user. Ultimately, a hybrid flight controller capable of handling the multi-modal locomotion cycle with high degrees of autonomy will be necessary for many applications.

Soft Robotics The field of soft robotics has flourished in the last decade. Scientists have realised the potential that compliant structures hold. Animals rely on tendons, muscles, membranes to move, which are all soft elements. Robots still cannot approach the grace and fluidity with which animals traverse terrain, and nature remains a potent resource for roboticists. Arguably, the defining performance difference between synthetic and natural systems is robustness - the ability to cope with change, perturbation and temporary locomotion deficits. In nature this is often accomplished with flexible structures that bend and adapt to change, absorb energy and protect from shock. The robots presented in this book mostly function with rigid elements, but aerial-aquatic robots are expected to be able to make use of soft technology in the near future. This will lead to improved transitions and safer use around humans.

Safety Building on the previous argument, the safety of aerial-aquatic robot is for now not sufficient for critical applications and can be vastly improved. The use of high pressures (as used in the AquaMAV and Consecutive Jump-glider robots) or propellers (as used in the AquaMAV, SailMAV and MEDUSA robots) can lead to dangerous situations in case of contact with humans or the environment. The applicability of lightweight aerial-aquatic robots will be higher if safer methods are developed. This will for instance include investigating novel materials, including soft materials, new propulsion methods, and more robust and reliable control approaches.

Robustness to environmental conditions As evident from this book, aerial-aquatic robots are intended for usage in outdoor environments, specifically in and around bodies of water. In addition to handling the habitual difficulties of outdoor operations, such as variable and often unpredictable wind and weather conditions, aerial-aquatic robots thus need to be robust to waves and water currents. Moreover, many of the specific applications envisaged for such robots Sect. 2.2, involve operation in close proximity to safety-critical infrastructure (e.g. offshore platforms) or delicate flora and fauna (e.g. coral reefs). While many of the solutions presented so far are highly promising, explicit consideration of realistic outdoor conditions will be essential before aerial-aquatic robots can be used in practice. This will involve exploring new control solutions, designs and materials, but will also require in-depth study of the environment itself, e.g. modelling the effect of waves and currents specifically on small-scale robots.

14.6 Concluding Remarks

Remarkable advances in robotics in recent years have given synthetic machines unprecedented access to the world around them. However, locomotion capabilities of robots are still far below that of animals, for example, and there remains a great deal of opportunity for further innovation in locomotion.

This innovation is needed, because the utility of any robot is fundamentally tied to its ability to traverse its environment. The advanced locomotion abilities that aerial-aquatic robots can deliver allow equally advanced data collection. This can enable us to address many important questions in both the natural world and the built environment.

But making a robot that can acquire this data is a huge challenge and requires creative thinking, judicious engineering and a willingness to get wet. This book shows the breadth of solutions that can be conceived of to move between sea and sky, and hopefully will be only the beginning of the story of aerial-aquatic locomotion.

Raph, Rob, Sophie and Mirko.

References

1. R. Siddall, M. Kov, Launching the AquaMAV: bioinspired design for aerial-aquatic robotic platforms. Bioinspir. Biomim. **9**, 31001–15 (2014)
2. R. Zufferey, A. Ortega Ancel, A. Farinha, R. Siddall, S.F. Armanini, M. Nasr, R.V. Brahmal, G. Kennedy, M. Kovac, Consecutive aquatic jump-gliding with water-reactive fuel. Sci. Robot. **4** (2019)
3. R. Zufferey, A. Ortega Ancel, C. Raposo, S.F. Armanini, A. Farinha, R. Siddall, I. Berasaluce, H. Zhu, M. Kovac, SailMAV: design and implementation of a novel multi-modal flying sailing robot. IEEE Robot. Autom. Lett. **4**(3), 1 (2019)
4. D. Debruyn, R. Zufferey, S.F. Armanini, C. Winston, A. Farinha, Y. Jin, M. Kovac, MEDUSA: a multi-environment dual-robot for underwater sample acquisition. IEEE Robot. Autom. Lett. **5**(3), 4564–4571 (2020)
5. R. Siddall, M. Kovac, Fast aquatic escape with a jet thruster. IEEE/ASME Trans. Mechatron. **PP**(99), 1 (2016)
6. R. Siddall, A. Ortega Ancel, M. Kovač, Wind and water tunnel testing of a morphing aquatic micro air vehicle. Interface Focus **7**(1) (2017)
7. S.F. Armanini, R.J. Siddall, M. Kovač, Modelling and simulation of a bioinspired aquatic micro aerial vehicle, in *AIAA Aviation 2019 Forum* (2019), pp. 1–18
8. Y.H. Tan, R. Siddall, M. Kovac, Efficient aerial-aquatic locomotion with a single propulsion system. IEEE Robot. Autom. Lett. **2**(3), 1304–1311 (2017)
9. M.A. Woodward, M. Sitti, MultiMo-Bat: a biologically inspired integrated jumping gliding robot. Int. J. Robot. Res. (2014)
10. R. Bachmann, F. Boria, R. Vaidyanathan, P. Ifju, R.D. Quinn, A biologically inspired micro-vehicle capable of aerial and terrestrial locomotion. Mech. Mach. Theory **44**, 513–526 (2009)
11. K. Peterson, P. Birkmeyer, R. Dudley, R.S. Fearing, A wing-assisted running robot and implications for avian flight evolution. Bioinspir. Biomim. **6**, 46008 (2011)
12. M. Kovač, The bioinspiration design paradigm: a perspective for soft robotics. Soft Robot. **1**(1), 28–37 (2014)
13. A. Ijspeert, A. Crespi, D. Ryczko, J. Cabelguen, From swimming to walking with a salamander robot driven by a spinal cord model. Science **315**, 20–1416 (2007)
14. S. Floyd, T. Keegan, J. Palmisano, M. Sitti, A novel water running robot inspired by basilisk lizards, in *IEEE/RSJ International Conference on Intelligent Robots and Systems* (2006), pp. 5430–5436
15. U. Saranli, M. Buehler, D.E. Koditschek, RHex: a simple and highly mobile hexapod robot. Int. J. Robot. Res. **20**, 616–631 (2001)

R. Zufferey et al., *Between Sea and Sky: Aerial Aquatic Locomotion in Miniature Robots*, Biosystems & Biorobotics 29, https://doi.org/10.1007/978-3-030-89575-4

16. K. Nagatani, S. Kiribayashi, Y. Okada, K. Otake, K. Yoshida, S. Tadokoro, T. Nishimura, T. Yoshida, E. Koyanagi, Emergency response to the nuclear accident at the Fukushima Daiichi nuclear power plants using mobile rescue robots. J. Field Robot. **30**(1), 44–63 (2013)
17. R. Murphy, E. Steimle, C. Griffin, C. Cullins, M. Hall, K. Pratt, Cooperative use of unmanned sea surface and micro aerial vehicles at Hurricane Wilma. J. Field Robot. **25**(3), 164–180 (2008)
18. P. Stevenson, Report on air launched autonomous underwater vehicles, National Oceanography Centre Research & Consultancy Report, vol. 44(04) (2011)
19. P. Stevenson, S.D. McPhail, M. Tsimplis, E. Higgins, Air launched platforms - a new approach for underwater vehicles, in *IEEE Oceans Conference* (2009), pp. 1–8
20. F. Shkurti, A. Xu, M. Meghjani, J.C. Gamboa Higuera, Y. Girdhar, P. Giguere, B.B. Dey, J. Li, A. Kalmbach, C. Prahacs, et al., Multi-domain monitoring of marine environments using a heterogeneous robot team, in *2012 IEEE/RSJ International Conference on Intelligent Robots and Systems (IROS)* (2012), pp. 1747–1753
21. W. Shyy, *Aerodynamics of Low Reynolds Number Flyers*, vol. 22 (Cambridge University Press, Cambridge, 2008)
22. D. Floreano, *Flying Insects and Robots* (Springer, Berlin, 2009)
23. W. Shyy, P. Ifju, D. Viieru, Membrane wing-based micro air vehicles. Appl. Mech. Rev. **58**, 283–301 (2005)
24. M.-S. Park, Y.-R. Jung, W.-G. Park, Numerical study of impact force and ricochet behavior of high speed water-entry bodies. Comput. Fluids **32**, 939–951 (2003)
25. J. Liang, X. Yang, T. Wang, G. Yao, W. Zhao, Design and experiment of a bionic gannet for plunge-diving. J. Bionic Eng. **10**(3), 282–291 (2013)
26. A. Shaw, A. Al-Shamma'a, Experimental investigations of electromagnetic wave propagation in seawater, in *Proceedings of the 36th European Microwave Conference* (2006), pp. 575–572
27. A. Gao, A.H. Techet, Design considerations for a robotic flying fish, in *Oceans 2011* (2011)
28. T.M. Williams, Swimming by sea otters: adaptations for low energetic cost locomotion. J. Comp. Physiol. **164**, 815–824 (1989)
29. V.M. Ortega-Jiménez, S. Álvarez-Borrego, S. Arriaga-Ramírez, M. Renner, E.S. Bridge, Takeoff flight performance and plumage wettability in Cassin's Auklet Ptychoramphus aleuticus, Xantus's Murrelet Synthliboramphus hypoleucus and Leach's Storm-petrel Oceanodroma leucorhoa. J. Ornithol. **151**, 169–177 (2009)
30. K. Kawachi, Y. Inada, A. Azuma, Optimal flight path of flying fish. J. Theor. Biol. **163**(2), 145–159 (1993)
31. J. Cook, N. Oreskes, P.T. Doran, W.R.L. Anderegg, B. Verheggen, E.W. Maibach, J.S. Carlton, S. Lewandowsky, A.G. Skuce, S.A. Green, D. Nuccitelli, P. Jacobs, M. Richardson, B. Winkler, R. Painting, K. Rice, Consensus on consensus: a synthesis of consensus estimates on human-caused global warming. Environ. Res. Lett. **11**(4) (2016)
32. M.W. Matthews, A current review of empirical procedures of remote sensing in inland and near-coastal transitional waters. Int. J. Remote Sens. **32**(21), 6855–6899 (2011)
33. S.C. Palmer, T. Kutser, P.D. Hunter, Remote sensing of inland waters: challenges, progress and future directions. Remote Sens. Environ. **157**, 1–8 (2015)
34. P.E. Land, H.S. Findlay, J.D. Shutler, I.G. Ashton, T. Holding, A. Grouazel, F. Girard-Ardhuin, N. Reul, J.F. Piolle, B. Chapron, Y. Quilfen, R.G. Bellerby, P. Bhadury, J. Salisbury, D. Vandemark, R. Sabia, Optimum satellite remote sensing of the marine carbonate system using empirical algorithms in the global ocean, the Greater Caribbean, the Amazon Plume and the Bay of Bengal. Remote Sens. Environ. **235**, 111469 (2019)
35. R. Morrow, L.-L. Fu, F. Ardhuin, M. Benkiran, B. Chapron, E. Cosme, F. d'Ovidio, J.T. Farrar, S.T. Gille, G. Lapeyre, P.-Y. Le Traon, A. Pascual, A. Ponte, B. Qiu, N. Rascle, C. Ubelmann, J. Wang, E.D. Zaron, Global observations of fine-scale ocean surface topography with the surface water and ocean topography (SWOT) mission. Front. Mar. Sci. **6**, 1–19 (2019)
36. L. Parker, Microplastics have moved into virtually every crevice on Earth
37. C. Li, R. Busquets, L.C. Campos, Assessment of microplastics in freshwater systems: a review. Sci. Total Environ. **707**, 135578 (2020)

38. European Centre for River Restoration (ECRR)
39. Y.M. Bar-On, R. Phillips, R. Milo, The biomass distribution on Earth, in *Proceedings of the National Academy of Sciences of the United States of America*, vol. 115(25) (2018)
40. R. Siddall, T. Fukushima, D. Bardhi, B. Perteshoni, A. Morina, E. Hasimja, Y. Dujaka, G. Haziri, L. Martin, H. Banerjee, A. Jusufi, Compliance, mass distribution and contact forces in cursorial and scansorial locomotion with biorobotic physical models. Adv. Robot. **35**(7) (2021)
41. R. Siddall, V. Ibanez, G. Byrnes, R.J. Full, A. Jusufi, Mechanisms for mid-air reorientation using tail rotation in gliding geckos. Integr. Comp. Biol. **61**(2) (2021)
42. T. Fukushima, R. Siddall, F. Schwab, S.L.D. Toussaint, G. Byrnes, J.A. Nyakatura, A. Jusufi, Inertial tail effects during righting of squirrels in unexpected falls: from behavior to robotics. Integr. Comp. Biol. **61**(2) (2021)
43. R. Dudley, G. Byrnes, S.P. Yanoviak, B. Borrell, R.M. Brown, J.A. McGuire, Gliding and the functional origins of flight: biomechanical novelty or necessity? Annu. Rev. Ecol. Evol. Syst. **38**(1), 179–201 (2007)
44. T. Aigeldinger, F. Fish, Hydroplaning by ducklings: overcoming limitations to swimming at the water surface. J. Exp. Biol. **198**, 1567–1574 (1995)
45. R.J. Lock, R. Vaidyanathan, S.C. Burgess, J. Loveless, Development of a biologically inspired multi-modal wing model for aerial-aquatic robotic vehicles through empirical and numerical modelling of the common guillemot, Uria aalge. Bioinspir. Biomim. **5**, 46001 (2010)
46. S. Garthe, S. Benvenuti, W.A. Montevecchi, Pursuit plunging by northern gannets (Sula bassana) feeding on capelin (Mallotus villosus). Proc. R. Soc. Lond. Ser. B: Biol. Sci. **267**(1454), 1717–1722 (2000)
47. P.Y. Daoust, G.V. Dobbin, Descriptive anatomy of the subcutaneous air diverticula in the northern gannet Morus bassanus. Seabird **21**, 64–76 (2008)
48. Y. Ropert-Coudert, D. Grémillet, P. Ryan, A. Kato, Y. Naito, Y. Le Maho, Between air and water: the plunge dive of the Cape Gannet Morus capensis. Ibis **146**(2), 281–290 (2004)
49. T.M. Wang, X.B. Yang, J.H. Liang, G.C. Yao, W.D. Zhao, CFD based investigation on the impact acceleration when a gannet impacts with water during plunge diving. Bioinspir. Biomim. **8**(3), 36006 (2013)
50. J. Davenport, R.N. Hughes, M. Shorten, P.S. Larsen, Drag reduction by air release promotes fast ascent in jumping emperor penguins - a novel hypothesis. Mar. Ecol. Prog. Ser. **430**, 171–182 (2011)
51. G. Iosilevskii, D. Weihs, Speed limits on swimming of fishes and cetaceans. J. R. Soc. Interface **5**(20), 329–338 (2008)
52. Y. Ropert-Coudert, F. Daunt, A. Kato, P.G. Ryan, S. Lewis, K. Kobayashi, Y. Mori, D. Grémillet, S. Wanless, Underwater wingbeats extend depth and duration of plunge dives in northern gannets Morus bassanus. J. Avian Biol. **40**(4), 380–387 (2009)
53. S. Wanless, J.D. Okill, Body measurements and flight performance of adult and juvenile gannets Morus bassanus. Ringing Migr. **15**(2), 101–103 (1994)
54. F. Richardson, Functional aspects of the pneumatic system of the California Brown Pelican. The Condor **41**(1), 13–17 (1939)
55. W.L. Morgan, Feeding methods of the short-tailed shearwater Puffinus tenuirostris. Emu **82**(4), 226–227 (1982)
56. J. Rayner, Pleuston: animals which move in water and air. Endeavour **10**(2), 58–64 (1986)
57. J.R. Lovvorn, D.R. Jones, Biomechanical conflicts between adaptations for diving and aerial flight in estuarine birds. Estuaries **17**(1), 62–75 (1994)
58. F. Quintana, R.P. Wilson, P. Yorio, Dive depth and plumage air in wettable birds: the extraordinary case of the imperial cormorant. Mar. Ecol. Prog. Ser. **334**, 299–310 (2007)
59. G. Ribak, D. Weihs, Z. Arad, How do cormorants counter buoyancy during submerged swimming? J. Exp. Biol. **207**(12), 2101–2114 (2004)
60. J.R. Lovvorn, G.A. Liggins, M.H. Borstad, S.M. Calisal, J. Mikkelsen, Hydrodynamic drag of diving birds: effects of body size, body shape and feathers at steady speeds. J. Exp. Biol. **204**, 1547–1557 (2001)

61. R.P. Wilson, K. Hustler, P.G. Ryan, A.E. Burger, E. Christian, S. Url, Diving birds in cold water: do Archimedes and Boyle determine energetic costs? Am. Nat. **140**(2), 179–200 (1992)
62. R.M. Sellers, Wing-spreading behaviour of the cormorant Phalacrocorax carbo. Ardea-Wageningen **83**, 27 (1995)
63. S. Sudo, T. Yano, Y. Kan, Y. Yamada, K. Tsuyuki, Swimming behavior of small diving beetles. J. Adv. Sci. **18**(1/2), 46–49 (2006)
64. I. Ribera, A.N. Nilsson, Morphometric patterns among diving beetles (Coleoptera: Noteridae, Hygrobiidae, and Dytiscidae). Can. J. Zool. **73**(12), 2343–2360 (1995)
65. G.T. Reels, R. Dow, Underwater oviposition behavior in two species of Euphaea in Borneo and Hong Kong (Odonata: Euphaeidae). Int. J. Odonatol. **9**(2), 197–204 (2006)
66. P.J. Mill, R.S. Pickard, Jet-propulsion in Anisopteran dragonfly larvae. J. Comp. Physiol. A **97**(4), 329–338 (1975)
67. M. Burrows, M.D. Picker, Jumping mechanisms and performance of pygmy mole crickets (Orthoptera, Tridactylidae). J. Exp. Biol. **213**(14), 2386–2398 (2010)
68. J. Davenport, How and why do flying fish fly?, Technical Report 2 (1994)
69. H. Park, H. Choi, Aerodynamic characteristics of flying fish in gliding flight. J. Exp. Biol. **213**(19), 3269–3279 (2010)
70. J. Davenport, Wing-loading, stability and morphometric relationships in flying fish (Exocoetidae) from the North-eastern Atlantic. J. Mar. Biol. Assoc. UK **72**, 25 (2009)
71. R.T. Jones, T.A. Lasinski, Effect of winglets on the induced drag of ideal wing shapes. NASA Tech. Memo. **81230**(81), 30 (1980)
72. W. Shyy, Y. Lian, J. Tang, D. Viieru, H. Liu, *Aerodynamics of Low Reynolds Number Flyers*, vol. 9780521882 (Cambridge University Press, Cambridge, 2007)
73. A. Azuma, *The Biokinetics of Flying and Swimming*. AIAA Education Series, 2nd edn. (American Institute of Aeronautics and Astronautics Inc, Reston, 2006)
74. K.L. Shephard, Functions for fish mucus (1994)
75. E.J. Anderson, M.E. DeMont, The locomotory function of the fins in the squid Loligo pealei. Mar. Freshw. Behav. Physiol. **38**(3), 169–189 (2005)
76. R.K. O'Dor, J.A. Hoar, D.M. Webber, F.G. Carey, Marine and freshwater behaviour and physiology squid (Loligo forbesi) performance and metabolic rates in nature (2013)
77. R.K. O'dor, The forces acting on swimming squid. J. Exp. Biol. **137**(1), 421–442 (1988)
78. K. Muramatsu, J. Yamamoto, T. Abe, K. Sekiguchi, N. Hoshi, Y. Sakurai, Oceanic squid do fly. Mar. Biol. **160**(5), 1171–1175 (2013)
79. R.W. Blake, Mechanics and physiology of animal swimming. J. Exp. Mar. Biol. Ecol. **191**(1), 131–132 (1995)
80. K.S. Cole, D.L. Gilbert, Jet propulsion of squid. Biol. Bull. **138**, 245–246 (1970)
81. S. Macia, New observations on airborne jet propulsion (flight) in squid, with a review of previous reports. J. Molluscan Stud. **70**(3), 297–299 (2004)
82. L.B. Spear, D.G. Ainley, Flight behaviour of seabirds in relation to wind direction and wing morphology. Ibis **139**, 221–233 (1997)
83. Y.Y. Watanabe, A. Takahashi, K. Sato, M. Viviant, C.A. Bost, Poor flight performance in deep-diving cormorants. J. Exp. Biol. **214**(3), 412–421 (2011)
84. T.R. Cook, G. Leblanc, Why is wing-spreading behaviour absent in blue-eyed shags? Anim. Behav. **74**, 649–652 (2007)
85. A. Muhammad, Q.V. Nguyen, H.C. Park, D.Y. Hwang, D. Byun, N.S. Goo, Improvement of artificial foldable wing models by mimicking the unfolding/folding mechanism of a beetle hind wing. J. Bionic Eng. **7**(2), 134–141 (2010)
86. K.H. Elliott, R.E. Ricklefs, A.J. Gaston, S.A. Hatch, J.R. Speakman, G.K. Davoren, High flight costs, but low dive costs, in auks support the biomechanical hypothesis for flightlessness in penguins. Proc. Natl. Acad. Sci. USA **110**(23), 9380–9384 (2013)
87. W. Chu, K. Lee, S. Song, M. Han, J. Lee, H. Kim, M. Kim, Y. Park, K. Cho, S. Ahn, Review of biomimetic underwater robots using smart actuators. Int. J. Precis. Eng. Manuf. **13**(7), 1281–1292 (2012)

88. Y.-H. Lin, R. Siddall, F. Schwab, T. Fukushima, H. Banerjee, Y. Baek, D. Vogt, Y.-L. Park, A. Jusufi, Modeling and control of a soft robotic fish with integrated soft sensing. Adv. Intell. Syst. (2021)
89. M. Kovac, M. Bendana, R. Krishnan, J. Burton, M. Smith, R. Wood, Multi-stage micro rockets for robotic insects. Robot.: Sci. Syst. VIII 185 (2013)
90. D.T. Grant, M. Abdulrahim, R. Lind, Design and analysis of biomimetic joints for morphing of micro air vehicles. Bioinspir. Biomim. **5**(4), 45007 (2010)
91. M. Abdulrahim, H. Garcia, G.F. Ivey, R. Lind, Flight testing a micro air vehicle using morphing for aeroservoelastic control, in *Collection of Technical Papers - AIAA/ASME/ASCE/AHS/ASC Structures, Structural Dynamics and Materials Conference*, vol. 3 (2004), pp. 1776–1792
92. M. Kovac, J.C. Zufferey, D. Floreano, Towards a self-deploying and gliding robot, *Flying Insects and Robots* (2009)
93. S.V. Shkarayev, P.G. Ifju, J.C. Kellogg, T.J. Mueller, *Introduction to the Design of Fixed-Wing Micro Air Vehicles Including Three Case Studies* (American Institute of Aeronautics and Astronautics, Reston, 2007)
94. R.J. Lock, S.C. Burgess, R. Vaidyanathan, Multi-modal locomotion: from animal to application. Bioinspir. Biomim. **9**(1), 11001 (2013)
95. K. Jones, F. Boria, R. Bachmann, R. Vaidyanathan, P. Ifju, R. Quinn, MMALV - The morphing micro air-land vehicle, in *IROS 2006* (2006)
96. L. Daler, Adaptive morphology for multi-modal locomotion (2015)
97. A. Vidyasagar, J.-C. Zufferey, D. Floreano, M. Kovac, Performance analysis of jump-gliding locomotion for miniature robotics. Bioinspir. Biomim. in press (2015)
98. K. Peterson, R.S. Fearing, Experimental dynamics of wing assisted running for a bipedal ornithopter, in *2011 IEEE/RSJ International Conference on Intelligent Robots and Systems (IROS)* (2011), pp. 5080–5086
99. K. Peterson, P. Birkmeyer, R. Dudley, R.S. Fearing, A wing-assisted running robot and implications for avian flight evolution. Bioinspir. Biomim. **6**, 46008 (2011)
100. M. Kovač, M. Fuchs, A miniature 7g jumping robot. Robot. Autom. (2008)
101. A.L. Desbiens, M.T. Pope, D.L. Christensen, E.W. Hawkes, M.R. Cutkosky, Design principles for efficient, repeated jumpgliding. Bioinspir. Biomim. **9**(2), 25009 (2014)
102. J.-P.W. Ore, Autonomous aerial water sampling. Ph.D. thesis (2014)
103. M. Schwarzbach, M. Laiacker, M. Mulero-Pazmany, K. Kondak, Remote water sampling using flying robots, in *2014 International Conference on Unmanned Aircraft Systems (ICUAS)* (2014), pp. 72–76
104. R. Eubank, E. Atkins, D. Macy, Autonomous guidance and control of the flying fish ocean surveillance platform, in *AIAA Infotech@ Aerospace Conference* (2009), pp. 1–12
105. J. Eastgate, R. Goddard, Submersible aircraft concept design study, Naval Surface Warfare Center Technical Report (2011)
106. B.D. Reid, *The Flying Submarine: The Story of the Invention of the Reid Flying Submarine, RFS-1*, Eagle edn. (2004)
107. H.-N. Nguyen, R. Siddall, B. Stephens, A. Navarro-Rubio, M. Kovac, A passively adaptive microspine grapple for robust, controllable perching, in *2019 2nd IEEE International Conference on Soft Robotics (RoboSoft)* (2019), pp. 80–87
108. M.M. Maia, P. Soni, F.J. Diez, Demonstration of an aerial and submersible vehicle capable of flight and underwater navigation with seamless air-water transition. CoRR (2015), arXiv:1507.01932
109. H. Alzu'bi, I. Mansour, O. Rawashdeh, Loon copter implementation of a hybrid unmanned aquatic-aerial quadcopter with active buoyancy control. J. Field Robot. (2017)
110. P. Drews, A.A. Neto, M. Campos, Hybrid unmanned aerial underwater vehicle: modeling and simulation, in *2014 IEEE/RSJ International Conference on Intelligent Robots and Systems (IROS 2014)* (IEEE, 2014), pp. 4637–4642
111. D. Lu, C. Xiong, Z. Zeng, L. Lian, A multimodal aerial underwater vehicle with extended endurance and capabilities, in *Proceedings - IEEE International Conference on Robotics and Automation*, vol. 2019 (2019), pp. 4674–4680

112. J. Moore, A. Fein, W. Setzler, Design and analysis of a fixed-wing unmanned aerial-aquatic vehicle, in *IEEE International Conference on Robotics and Automation (ICRA)* (2018), pp. 1236–1243
113. W. Stewart, W. Weisler, M. Macleod, T. Powers, A. Defreitas, R. Gritter, M. Anderson, K. Peters, A. Gopalarathnam, M. Bryant, Design and demonstration of a seabird-inspired fixed-wing hybrid UAV-UUV system. Bioinspir. Biomim. **13**(5) (2018)
114. A. Fabian, Y.F. Feng, E. Swartz, D. Turmer, R. Wang, Hybrid aerial underwater vehicle (2012)
115. T.M. Wang, X.B. Yang, J.H. Liang, G.C. Yao, W.D. Zhao, CFD based investigation on the impact acceleration when a gannet impacts with water during plunge diving. Bioinspir. Biomim. **8**(3), 36006 (2013)
116. F.M. Rockenbauer, S.L. Jeger, L. Beltran, M.A. Berger, M. Harms, N. Kaufmann, M. Rauch, M. Reinders, N. Lawrance, T. Stastny, R.Y. Siegwart, Dipper: a dynamically transitioning aerial-aquatic unmanned vehicle. Robot.: Sci. Syst. (2021)
117. R.A. Peloquin, D. Thibault, A.L. Desbiens, Design of a passive vertical takeoff and landing aquatic UAV. IEEE Robot. Autom. Lett. **2**(2), 381–388 (2017)
118. E. Tetreault, D. Rancourt, A.L. Desbiens, Active vertical takeoff of an aquatic UAV. IEEE Robot. Autom. Lett. **5**(3), 4844–4851 (2020)
119. D. Mackenzie, A flapping of wings. Science **335**(6075), 1430–1433 (2012)
120. K.W. Moored, F.E. Fish, T.H. Kemp, H. Bart-Smith, Batoid fishes: inspiration for the next generation of underwater robots. Mar. Technol. Soc. J. **45**, 99–109 (2011)
121. J.S. Izraelevitz, M.S. Triantafyllou, A novel degree of freedom in flapping wings shows promise for a dual aerial/aquatic vehicle propulsor (2014), arXiv:1412.3843
122. Y. Chen, E.F. Helbling, N. Gravish, K. Ma, R.J. Wood, Hybrid aerial and aquatic locomotion in an at-scale robotic insect, in *2015 IEEE/RSJ International Conference on Intelligent Robots and Systems (IROS)* (2015)
123. W. Shyy, Y. Lian, J. Tang, D. Viieru, H. Liu, *Aerodynamics of Low Reynolds Number Flyers* (Cambridge University Press, Cambridge, 2007)
124. A. Azuma, *The Biokinetics of Flying and Swimming* (American Institute of Aeronautics and Astronautics, Inc., Reston, 2006)
125. UIUC applied aerodynamics group, UIUC Airfoil Coordinates Database
126. Airfoil Tools
127. D. Savitsky, Hydrodynamic design of planing hulls. Mar. Technol. **1**(1) (1964)
128. P. Timmerman, J.P. van der Weele, On the rise and fall of a ball with linear or quadratic drag. Am. J. Phys. **67**(6), 538–546 (1999)
129. T. Von Karman, The impact on seaplane floats during landing (1929)
130. B. Chang, M. Croson, L. Straker, S. Gart, C. Dove, J. Gerwin, S. Jung, How seabirds plunge-dive without injuries, in *Proceedings of the National Academy of Sciences* (2016), p. 201608628
131. F.E. Fish, Influence of hydrodynamic-design and propulsive mode on mammalian swimming energetics. Aust. J. Zool. **42**(1), 79–101 (1994)
132. C.J. Gommes, A more thorough analysis of water rockets: moist adiabats, transient flows, and inertial forces in a soda bottle. Am. J. Phys. **78**, 236 (2010)
133. D. I. N. E N, 13445-3, *Unfired Pressure Vessels*, vol. 3 (2005)
134. C. Wang, *Applied Elasticity* (McGraw-Hill, New York, 1953)
135. E.C. Polhamus, *A Concept of the Vortex Lift of Sharp-Edge Delta Wings Based on a Leading-Edge-Suction Analogy* (National Aeronautics and Space Administration, Springfield, 1966)
136. S.F. Hoerner, Fluid-dynamic drag (1965)
137. M. Shiffman, D.C. Spencer, The force of impact on a cone striking a water surface (vertical entry). Commun. Pure Appl. Math. **4**(4), 379–417 (1951)
138. J. Rayner, Pleuston: animals which move in water and air. Endeavour **10**(2), 58–64 (1986)
139. R. Siddall, M. Kovač, Launching the AquaMAV: bioinspired design for aerial-aquatic robotic platforms. Bioinspir. Biomim. **9**(3), 31001 (2014)
140. J. Yuh, M. West, Underwater robotics. Adv. Robot. **15**(5), 609–639 (2001)

141. J.-P. Ore, S. Elbaum, A. Burgin, B. Zhao, C. Detweiler, Autonomous aerial water sampling, in *The 9th International Conference on Field and Service Robots (FSR)* (2013)
142. S. Barbarino, O. Bilgen, R.M. Ajaj, M.I. Friswell, D.J. Inman, A review of morphing aircraft. J. Intell. Mater. Syst. Struct. **22**, 823–877 (2011)
143. J.D. Jacob, S.W. Smith, Design limitations of deployable wings for small low altitude UAVs, in *47th AIAA Aerospace Sciences Meeting Including the New Horizons Forum and Aerospace Exposition* (2009), pp. 5–8
144. L. Dufour, K. Owen, S. Mintchev, D. Floreano, A drone with insect-inspired folding wings (preprint), in *2016 IEEE/RSJ International Conference on Intelligent Robots and Systems (IROS 2016)* (2016)
145. J. Hall, K. Mohseni, D. Lawrence, P. Geuzaine, Investigation of variable wing-sweep for applications in micro air vehicles, in *Infotech Aerospace* (2005), pp. 26–29
146. M. Kovač, J.-C. Zufferey, D. Floreano, Towards a self-deploying and gliding robot, *Flying Insects and Robots* (Springer, Berlin, 2010), pp. 271–284
147. T.W. Cronin, Visual optics: accommodation in a splash. Curr. Biol. **22**(20), R871–R873 (2012)
148. S. Jung, B. Chang, M. Croson, L. Straker, C. Dove, Diving dynamics of seabirds, in *APS March Meeting Abstracts* (2015)
149. M. Okamoto, A. Azuma, Aerodynamic characteristics at low Reynolds number for wings of various planforms. AIAA J. **49**(6), 1135–1150 (2011)
150. W. Shyy, P. Ifju, D. Viieru, Membrane wing-based micro air vehicles. Appl. Mech. Rev. **58**(4), 283 (2005)
151. M. Drela, QPROP: propeller/windmill analysis and design (2007)
152. E.E. Larrabee, *Practical Design of Minimum Induced Loss Propellers* (SAE, Warrendale, 1979)
153. A. Pope, W.H. Rae, *Low-Speed Wind Tunnel Testing* (Wiley, New York, 1984)
154. T.J. Mueller, *Fixed and Flapping Wing Aerodynamics for Micro Air Vehicle Applications* (American Institute of Aeronautics & Astronautics, Reston, 2001)
155. V.A. Tucker, The energetic cost of moving about: walking and running are extremely inefficient forms of locomotion. Much greater efficiency is achieved by birds, fish—and bicyclists. Am. Sci. **63**(4), 413–419 (1975)
156. R.-A. Peloquin, D. Thibault, A.L. Desbiens, Design of a passive vertical takeoff and landing aquatic UAV. IEEE Robot. Autom. Lett. **2**(2), 381–388 (2017)
157. T.P. Krone, Y.S. Baik, J.M. Rausch, L.P. Bernal, Aerodynamics of a successful perching maneuver, in *AIAA Paper*, vol. 218 (2011)
158. Y. Ropert-Coudert, D. Grémillet, P. Ryan, A. Kato, Y. Naito, Y. Le Maho, Between air and water: the plunge dive of the Cape Gannet Morus capensis. Ibis **146**(2), 281–290 (2004)
159. A. Vilches, J. Arizaga, I. Salvo, R. Miranda, An experimental evaluation of the influence of water depth and bottom color on the common kingfisher's foraging performance. Behav. Process. **98**, 25–30 (2013)
160. M.M. Maia, P. Soni, F.J. Diez, Demonstration of an aerial and submersible vehicle capable of flight and underwater navigation with seamless air-water transition (2015), arXiv:1507.01932
161. C.C. de Visser, J.A. Mulder, Q.P. Chu, Global nonlinear aerodynamic model identification with multivariate splines, in *AIAA Atmospheric Flight Mechanics Conference*, vol. 5058 (2009)
162. R.C. Nelson, Flight stability and automatic control (1998), p. 441
163. J. Yang, J. Feng, Y. Li, A. Liu, J. Hu, Z. Ma, Water-exit process modeling and added-mass calculation of the submarine-launched missile. Pol. Marit. Res. **24** (2017)
164. V. Mishra, S. Vengadesan, Translational added mass of axisymmetric underwater vehicles with forward speed using computational fluid dynamics. J. Ship Res. **55** (2011)
165. D. Perrault, N. Bose, S. O'Young, C.D. Williams, Sensitivity of AUV added mass coefficients to variations in hull and control plane geometry. Ocean Eng. **30** (2003)
166. C.E. Brennen, Review of added mass and fluid inertial forces. Technical report, Naval Civil Engineering Laboratory (1982)
167. A.I. Korotkin, *Added Masses of Ship Structures*, 1st edn. (Springer, Netherlands, 2009)

168. H. Wagner, Über die Entstehung des dynamischen Auftriebes von Tragflügeln. ZAMM - Zeitschrift für Angewandte Mathematik und Mechanik **5**(1) (1925)
169. M. Shiffman, D.C. Spencer, The force of impact on a cone striking a water surface (vertical entry). Commun. Pure Appl. Math. **4**(4), 379–417 (1951)
170. Y.M. Scolan, A.A. Korobkin, Water entry of a body which moves in more than six degrees of freedom, in *Proceedings of the Royal Society A: Mathematical, Physical and Engineering Sciences*, vol. 471(2177) (2015)
171. R. Siddall, A. Ortega Ancel, M. Kovac, Wind and water tunnel testing of a morphing aquatic micro air vehicle. Interface Focus
172. J.S. Koh, E. Yang, G.P. Jung, S.P. Jung, J.H. Son, S.I. Lee, P.G. Jablonski, R.J. Wood, H.Y. Kim, K.J. Cho, Jumping on water: surface tension-dominated jumping of water striders and robotic insects. Science **349**(6247), 517–521 (2015)
173. B. Chang, J. Myeong, E. Virot, C. Clanet, H.Y. Kim, S. Jung, Jumping dynamics of aquatic animals. J. R. Soc. Interface **16**(152) (2019)
174. L. Wen, T. Wang, G. Wu, J. Liang, Quantitative thrust efficiency of a self-propulsive robotic fish: experimental method and hydrodynamic investigation. IEEE/ASME Trans. Mechatron. **18**(3), 1027–1038 (2013)
175. J. Mo, Z. Miao, B. Li, Y. Zhang, Z. Song, Design, analysis, and performance verification of a water jet thruster for amphibious jumping robot. Proc. Inst. Mech. Eng. Part C: J. Mech. Eng. Sci. (2019)
176. L. Jian, Z. Jianing, W. Zhenlong, CFD simulation of effect of vortex ring for squid jet propulsion and experiments on a bionic jet propulsor. Int. J. u- e- Serv. Sci. Technol. **9**(5), 211–226 (2016)
177. F.G. Serchi, A. Arienti, I. Baldoli, C. Laschi, An elastic pulsed-jet thruster for soft unmanned underwater vehicles, in *Proceedings - IEEE International Conference on Robotics and Automation* (2013), pp. 5103–5110
178. W.A. Churaman, L.J. Currano, C.J. Morris, J.E. Rajkowski, S. Bergbreiter, The first launch of an autonomous thrust-driven microrobot using nanoporous energetic silicon. J. Microelectromech. Syst. **21**(1), 198–205 (2012)
179. Y. Chen, H. Wang, E.F. Helbling, N.T.N. Jafferis, R. Zufferey, A. Ong, K. Ma, N. Gravish, P. Chirarattananon, M. Kovac, R.J.R. Wood, A biologically inspired, flapping-wing, hybrid aerial-aquatic microrobot. Sci. Robot. **2**(11), 1–12 (2017)
180. M.T. Tolley, R.F. Shepherd, M. Karpelson, N.W. Bartlett, K.C. Galloway, M. Wehner, R. Nunes, G.M. Whitesides, R.J. Wood, An untethered jumping soft robot, in *IEEE International Conference on Intelligent Robots and Systems* (IEEE, 2014), pp. 561–566
181. M. Loepfe, C.M. Schumacher, U.B. Lustenberger, W.J. Stark, An untethered, jumping roly-poly soft robot driven by combustion. Soft Robot. **2**(1), 33–41 (2015)
182. H. Newhouse, P. Payne, Underwater power source study, Technical report, DTIC Document (1981)
183. J. Borchsenius, S. Pinder, Underwater glider propulsion using chemical hydrides, in *OCEANS 2010 IEEE-Sydney* (2010), pp. 1–8
184. R. Siddall, G. Kennedy, M. Kovac, Explosive propulsion strategies for aquatic take off in robotics, in *2015 International Symposium on Robotics Research (ISRR)* (2015)
185. D. Roemmich, G. Johnson, S. Riser, R. Davis, J. Gilson, W.B. Owens, S. Garzoli, C. Schmid, M. Ignaszewski, The Argo program: observing the global oceans with profiling floats. Oceanography **22**(2), 34–43 (2011)
186. C.C. Eriksen, T.J. Osse, R.D. Light, T. Wen, T.W. Lehman, P.L. Sabin, J.W. Ballard, A.M. Chiodi, Seaglider: a long-range autonomous underwater vehicle for oceanographic research. IEEE J. Ocean. Eng. **26**(4), 424–436 (2001)
187. P. Voosen, Saildrone fleet could help replace aging buoys: robotic test in Pacific Ocean heralds new way to monitor El Nino events (2018)
188. C. Mordy, E. Cokelet, A. De Robertis, R. Jenkins, C. Kuhn, N. Lawrence-Slavas, C. Berchok, J. Crance, J. Sterling, J. Cross, P. Stabeno, C. Meinig, H. Tabisola, W. Burgess, I. Wangen, Advances in ecosystem research: Saildrone surveys of oceanography, fish, and marine mammals in the Bering Sea. Oceanography **30**(2) (2017)

189. M. Osborne, How do these boats sail faster than the wind? — KQED science (2013)
190. E.D. Cokelet, C. Meinig, N. Lawrence-Slavas, R. Jenkins, C.W. Mordy, H.M. Tabisola, J.N. Cross, The use of Saildrones to examine spring conditions in the Bering Sea: instrument comparisons, sea-ice meltwater, and Yukon River plume studies, in *OCEANS 2015 - MTS/IEEE Washington* (2015), pp. 1–7
191. I.S. Silva, F. Campopiano, G.S.V. Lopes, A.K. Uenojo, H.T. Silva, E.L. Pellini, A.A. Alvarez, E.A. Barros, Development of a trimaran ASV. IFAC-PapersOnLine **51**(29), 8–13 (2018)
192. A. Benedetto, P. Luca, M. Marco, A. Corrieri, D. Masti, L. Vanni, Development of an innovative and sustainable sail-drone, in *Conference Proceedings - 2017 17th IEEE International Conference on Environment and Electrical Engineering and 2017 1st IEEE Industrial and Commercial Power Systems Europe, EEEIC/I and CPS Europe 2017* (2017), pp. 1–6
193. I. Dathe, M. De Leo, Hydrodynamic characteristics of seaplanes as affected by hull shape parameters, in *Advanced Marine Vehicles Conference* (1989)
194. H. Furukawa, A.W. Blakeley, R.G.J. Flay, P.J. Richards, Performance of wing sail with multi element by two-dimensional wind tunnel investigations. JSME J. Fluid Sci. Technol. **10**(2) (2015)
195. A.W. Blakeley, R.G.J. Flay, P.J. Richards, Design and optimisation of multi-element wing sails for multihull yachts, in *Proceedings of the 18th Australasian Fluid Mechanics Conference, AFMC 2012* (2012)
196. G.H. Elkaim, Autonomous surface vehicle free-rotating wing sail section design and configuration analysis. J. Aircr. **45**(6), 1835–1852 (2008)
197. L. Jaulin, F. Le Bars, A simple controller for line following of sailboats. Technical report (2013)
198. C. Viel, U. Vautier, J. Wan, L. Jaulin, Position keeping control of an autonomous sailboat. IFAC-PapersOnLine **51**(29), 14–19 (2018)
199. J.-P. Ore, S. Elbaum, A. Burgin, C. Detweiler, Autonomous aerial water sampling. J. Field Robot. **32**(8), 1095–1113 (2015)
200. J.E. Adkins, R.S. Rivlin, Large elastic deformations of isotropic materials IX. The deformation of thin shells. Philos. Trans. R. Soc. Lond. Ser. A, Math. Phys. Sci. **244**(888), 505–531 (1952)
201. W.H. Yang, W.W. Feng, On axisymmetrical deformations of nonlinear membranes (1970)
202. D.D. Washington, Efficient methods for analyzing thin membranes subjected to transverse pressure and undergoing large deformation. The University of Alabama in Huntsville (2004)
203. Y.-Y. Hsu, K. Lucas, D. Davis, B. Elolampi, R. Ghaffari, C. Rafferty, K. Dowling, Novel strain relief design for multilayer thin film stretchable interconnects. IEEE Trans. Electron Devices **60**(7), 2338–2345 (2013)
204. S. McAllister, J.Y. Chen, A.C. Fernandez-Pello, *Fundamentals of Combustion Processes*. Mechanical Engineering Series (Springer, Berlin, 2011)